AF879771

Techies, Trust, and Trillionaires

How Blockchain and Emerging Technologies
Create Money for Africa

By Bert Kastel

Techies, Trust, and Trillionaires

By Bert Kastel

Copyright (C) 2018 by Berthold F. Kastel

All rights reserved.
No part of this publication may be reproduced, scanned, or distributed in any printed or electronic form without permission.

Published in the United States by Competitive Edge International, Inc. of Sarasota, Florida (USA).

Paperback ISBN 978-0-9741808-7-8

eBook ISBN 978-0-9741808-8-5

"If you always do what you've always done, you'll always get what you've always got."

-- Henry Ford --

Prelude

WHEN YOU ARRIVE at Telecom House in Kigali, as in many other African innovation hubs, you first go through security. The guards will take your ID, sign you in, and point you in the right direction. Then you go up an elevator. As the door opens, you see them. Young, educated Africans.

They are everywhere, sitting behind laptops, working alone or in small groups, discussing ideas, learning, designing, planning, creating, and building. The energy is palpable. At some innovation hubs there are just a handful of young women and men, but like here at kLab, often there are hundreds.

The current scientific consensus is that humans evolved in Africa, more specifically East Africa. In that respect, Africa is the oldest continent. But the exact world region where the oldest known humans lived is full of young people. Based on median age, Africa really is the youngest continent.

Its innovation hubs are filled with know-how, skills, creativity, ideas – and lots of drive and entrepreneurial energy. These are the brains of Africa, ready to take on the world, to succeed in life, to change their societies, transform, transcend, leapfrog. They don't like the status quo.

These are the prospective and actual **Heroes of Africa**.

* * *

Techies, Trust, and Trillionaires is not about Africa. It is not about Blockchain either. It is not even about money.

This book is about the potential of technology to transform and advance human societies anywhere. It also is about the need to actively engage and imagine and shape a future that uses emerging technologies to benefit *all of us*. If we do not do that, others will do so in a way that benefits *them*.

Although I spent months traveling and working with Africans in several countries, I don't claim to be an expert in Africa, nor even knowing whether there really can be one. What I understand, though, is that using Blockchain-technologies to integrate and empower individuals in Africa and all over the world can drive spectacular economic progress and social advances.

Therefore, I wrote this book considering and honoring the large number of young African minds, reflecting the world's greatest asset of driven, adaptive, and increasingly educated young brains. One-fifth of the world's young people live in Africa.

Tapping into these brains and unleashing them will not just be transformative for Africa but also become one of the biggest boons for the rest of the world. Who can afford to neglect the output of 800 million creative minds?

Therefore, I tell a tale of these many actual and prospective, likely and unlikely, known and unknown heroes of Africa. Those believing in the transformative and transcending power of technology, and particularly those working to use it.

* * *

The **Heroes of Africa** are entrepreneurs, technology enthusiasts, thought leaders, engineers, students, educators, politicians, business people, professionals, government employees, and social activists eager to pursue their dreams and aspirations while transforming their societies and thus a large continent and eventually the world. Blockchain is a big part of their game, but their ambitions go far beyond it: they are hell-bent on using any and all emerging technologies to catapult Africa into a shiny future.

Many are visionaries who see the bright potential of the continent and its likely important contributions to master humanity's challenges. Others are merely working on building a better future for themselves and their loved ones.

Not all of them are Africans, although most are. Some also are outsiders, "Muzungus." They have a lot of compassion for Africans, or an interest in partnering, working, and doing business with dynamic people and societies.

And so, this book became a story in three acts.

* * *

Act I describes the good fight of these heroes who are about to transform a continent. They are ambitious, creative, smart, intelligent, and always working hard. But they continuously run into limitations that keep them at a severe disadvantage, illustrated by an ever-present shortage of money. All the same, Blockchains and other emerging technologies promise a brighter future via new economic and investment models.

Act II analyzes the specific constraints that Africa's heroes encounter. Vicious cycles limit chances of success. The lack of even basic infrastructure and the fragmentation of already small economies leave few options. It becomes difficult to overcome Africa's specific historic and political limitations, and its geographical and economic barriers. But our heroes can use Blockchain technologies to change the trajectory and trigger massive investments that open the door to unleash African brains. Can they really cut through that web holding them down, the Gordian Knot standing in the way of their success?

Act III is about that future. It describes in more detail how the heroes of Africa can leverage Blockchains and other emerging technologies and effectively create value and money in a way that not just transforms Africa, but transcends it and possibly the world. The obstacles seem almost insurmountable, but with smarts and courage they can be overcome. When looking at the future, unhindered by misguided and limiting visions of the past, a possible path becomes visible, maybe even likely and obvious.

Can Africa's heroes find the courage to go down that path? Can they support each other and find the partners and Muzungu sidekicks they need to shape Africa's bright future?

Will they succeed, and unleash one of the greatest superpowers the world has ever seen – 800 million young, creative, hungry, ambitious, hustling African brains?

Let's see how the story ends.

Table of Contents

Act I

Blockchain and Africa's Leapfrogging Heroes

* * *

Chapter 1

Beyond the Echo Chamber

ON APRIL 4, 2018, I was sitting with over 2,000 people in the Walkerhill Hotel in Seoul, South Korea, to follow a panel of five leading thinkers and doers in the Blockchain world. I call them the "Fab Five" of Ethereum: Vitalik Buterin, Joseph Poon, Karl Floersch, Vlad Zamfir, and Justin Drake.

It was a really high-value and "cool" panel, with all five participants being likable, and highly intelligent, proven to be creative thinkers and developers. All have demonstrated to be effective doers also.

Something was amiss, though. No Africans.

Preceding this panel discussion, Aya Miyaguchi from the Ethereum Foundation had powerfully talked about her personal Blockchain story to seek inclusiveness and the empowerment of those in need, particularly women and people in developing nations.[1] But despite their difference in personalities, all five panelists seemed so – alike. I saw five 20 to 30-year-old males who all grew up in a rich world, all millionaires many times over, with similar jokes, demeanor, and rockstar-like status.

This was not specific to Ethereum. I mention the above group since the general perception of them, which I share, is very positive. They are open-minded and idealistic, likable and intelligent. There is nothing nefarious about them in that.

But Ethereum's Fab Five are in many ways typical of the Blockchain, or crypto, community. Of the 48 people featured on the

Deconomy web page,[2] there were zero Africans. Ian Grigg was the closest Africa came to be represented, since he works a lot in Africa, developing African solutions with Africans for Africans. Though, Ian is not really a native of that continent or of African descent.

Deconomy 2018 in Seoul reflected diversity of a kind, and it was one of the most impressive and most open-minded and high-quality Blockchain event I had attended in about a year. Still – Africa and the mainstream Blockchain world were missing out big time.

Talking to Oneself

Much of what is considered the "Blockchain community" is actually an echo chamber. There is so much idealism, talk about reaching out to billions, diversity, empowerment, financial inclusion, eliminating intermediaries, eradicating corruption, and leveling the playing field via a decentralized economy.

In reality, though, most and definitely the best-represented Blockchain conferences of 2016, 2017, and even 2018 took place in Hong Kong, Seoul, San Francisco, New York City, London, Berlin, Singapore, with the occasional Dubai, Sydney or Cancun sprinkled in. You can see the same people in all these places, talking to each other. In other words, it's an echo chamber.

As of 2018, most crypto-solutions are not built by or for the poor. Economics represented by Gross Domestic Product (GDP) and investment power still dominate the picture. Without diminishing the great efforts of many individuals and businesses, it still is almost all about Europe, East Asia, and North America.

I get it: Africa is a small market. It is also a difficult one and splintered into 55 economies. But this does not need to matter much. The opportunity in Africa is not the generation of short-term revenues and profits. It is the access to large numbers of young and creative minds. According to data from the United Nations (UN), compiled by Euromonitor International,[3] in 2012 China had 497 million people below age 30. India had 704 million. Africa had around 800 million, 70% of its population of about 1.2 billion.

It may be theoretically possible that almost all the best Blockchain brains are Caucasian or Eurasian males from rich countries, roughly between the age of 20 and 30. It just is extremely implausible. And therefore, it is at our own peril if we do not inspire, include and tap into at least some of the millions of African brains that want and need to contribute.

My point here is not to accuse anybody of direct or even indirect bigotry. Rather I am saying that we all are missing out here: We are wasting access to one of the greatest sources of the most important ingredient of human advancement in tomorrow's technology-driven world: brain power. More particularly we are not including the world's young people in areas that seem so predestined to contribute to, and gain from, Blockchain knowledge.

Blockchain in Africa: The Good, The Ugly, and The Real

African enthusiasm for crypto-technology is deep and widespread. At the end of 2017, African countries Nigeria (3rd) and Kenya (5th) had among the highest Bitcoin holding rates in the world,[4] as compared to their relative share of GDP. This was over ten times the adoption rate of the United States. Within just several months since the end of 2017, sporadic meet-ups spread across Africa and morphed into ever-bigger events. Large Blockchain organizations and some cryptocurrency foundations representing billion-dollar valuations are increasingly engaging on the continent.

This latter fact, while laudable, also comes with occasional not-so-pretty side effects. Sometimes, good intentions are not enough to prevent repeating unfortunate historic patterns. As so often, they have to do with money, or rather Africa's shortage of it.

During the organization of our Blockchain double-event (*TheWorkshop* and *Blockchain Hub*) at the 2018 Transform Africa Summit (TAS2018), we were able to first-hand observe several instances of this. While expressing support for the general cause of building a larger Blockchain community on the continent, some high-profile Blockchain organizations played hardball. They used the power of the purse in a questionable manner. We were effectively given what I sometimes call the "African Treatment."

This works by leveraging what in the rich world would be considered a relatively small amount of money and dangling it in front of people who need it, trying to get maximum leverage out of the process. In our case there were verbal promises in form of speaking participation and sponsorship intentions to organizers, followed by gradually reduced amounts. Eventually some were even asking for free rides or even payment (!) for us giving them market exposure and the chance to build mindshare. Such hard-nosed tactics are not unheard of in a cutthroat world of competitive businesses. In some ways this makes them "classical." However, they certainly are not "classy" because they exploit the particular dearth of financial alternatives in Africa, for an event that had no other funding since entrance to our workshop was free.

Of course, my African co-organizer was livid. This would hardly have been tried at comparable events in the rich world. It was possible for one reason only: there simply is no money in Africa. This then immediately distorts any business case and establishes an uneven power-relationship. It also makes it very difficult for Africans to compete in the world, even if in theory the only other resource needed is brainpower.

It comes down to money.

Chapter 2

It's All About Money

ROGER VER does not mince words.

> "More babies are dying in countries around the world because they have less economic freedom, more people are starving to death because they have less economic freedom, people are literally dying because of this."[5]

As Chief Executive Officer (CEO) of Bitcoin.com, one of the first adopters of Bitcoin, and the first investor in the Bitcoin and Blockchain space, Roger has been a major driving force in building the crypto and Blockchain community and ecosystem to grow to a market capitalization of hundreds of billions of dollars.

He followed up on the above by stating, "this is a life-and-death matter around the world," and challenging a large part of the Blockchain community's focus on "some science project [...] for a bunch of people living in First World countries with a bunch of money already, [without] the ability to help people that actually need the help the most."[6] Roger wants "to see Bitcoin Cash be available for every single person on the Planet regardless of how much money they have or where they are born."

Whenever I talk to Roger or listen to him, he shows this passion. To him, Blockchain in general and Bitcoin specifically are about "money" as a means to bring economic advancement to the masses in the world. In my personal experience, this extreme and idealistic focus on the core philosophical driving force of Bitcoin and Blockchain sets the "old guard" and many of the entrepreneurial leaders apart from most others. After all, most of the crypto-community seems to either enjoy never-ending debates about theoretical concepts, or ditch all deep thinking and just chase quick profits with initial coin offerings (ICOs). While many of them have followers and "haters," including Roger Ver, Vitalik Buterin, Dan Larimer, Charles Hoskinson, Craig Wright, Jed McCaleb, and many more, at least formally they all seem to pursue larger and philosophically motivated visions.

Talk is easy. In Africa, though, the "rubber hits the road." Several lessons directly apply to Blockchain in Africa, and to African competitiveness in the world of technology.

- For Africans today, due to a lack of money, life usually is about *surviving, not thriving.*
- For businesses operating in Africa today, low purchasing power and small economies mean *little revenue potential.*
- For the world outside Africa, engaging with Africa is *mostly a social effort and not so much a business investment.*

And there goes the respect. And the opportunity.

Digging deeper into what is perpetuating this cycle, it quickly becomes obvious that a profound lack of expensive infrastructure is the main culprit. As I will argue in more detail later, it likely is the main reason that holds back economic contributions of hundreds of millions of otherwise perfectly capable minds and brains.

The case for new Blockchain-enabled crypto-economic motivations and solutions could hardly be made more powerfully. *Money* is needed to unleash African brainpower. Money in the form of investments that make sense, are profitable and show a high return. Money in the form of payments for African products and services. Money that is stable enough to conduct commerce.

So, for all the likely and prospective heroes of Africa, breaking out of these severe limitations requires profoundly new economic models and business cases. This is what Blockchain is supposed to be best at, putting new concepts, ideas, and emerging technologies on steroids, and realizing them in a non-traditional, diversified, decentralized, bottom-up and community-based way.

African Blockchain Beginnings

In 2018 we can already see fledgling Blockchain beginnings all over Africa, although mostly related to social disruption and financial technologies (Fintech). Africa has several crypto marketplaces and regional Blockchain startups, and many initiatives, like Wala,[7] BitPesa,[8] Chamapesa,[9] or Eversend.[10] With mPesa and similar mobile money solutions, a part of the continent even has somewhat of a technological leadership role that in Fintech possibly is only matched by China. This is in a field that relates to one of Blockchain's biggest promises, a new form of cash system that eliminates the power of banks. There also are dozens of technology hubs with vibrant communities, the biggest of which are in Lagos, Nairobi, Cairo, Accra, and Johannesburg, and a variety of hundreds of innovation centers.

Despite all this, on a global scale African technology businesses are still minor actors. The Blockchain world like the general technology world is mostly led and dominated by others. It therefore remains important to keep engaging with each other and intensifying the conversations and cooperation. African's goal cannot be to develop Africa-only Blockchain technologies. After all, Blockchain is based on universal mathematics and cryptography anyway.

But Africa needs applications that directly address the problems of their societies, summarized in my three bullet points above, and even more such that leverage Africa's competitive advantage (the brains!). I will go much more in detail about prospective solutions, and how tokens and crypto-economics can help. It will not be by merely throwing out a coin, using buzzwords, and shouting "ICO," so much has to be clear.

Luckily, most of the global Blockchain community projects have an idealistic image to change the world by empowering individuals in a bottom-up manner. Considering the large number of Blockchain organizations talking about financial inclusion, though, the share of those with a strong footprint in Africa is still minuscule.

Some smaller organizations are present, like my own Tech844 and Healid, which engaged in 2017. Bigger players are active also. Most have been around longer and with more substantive investments. EOS is well represented in Kenya, Stellar shows flag big time across the continent, and Cardano has been active particularly in Ethiopia. Consensys[11] and the Blockchain for Social Impact Coalition (BSIC)[12] pursue multiple ventures in several countries, and Bitcoin Cash's nChain[13] showed up big time to the 2018 Transform Africa Summit and even committed to a substantial sponsorship.

As of mid-2018, though, even such well-known and outspoken altruistic organizations like the Ethereum Foundation or Ethereum's lead developers have shown little activity in the heart of the continent. Considering that context, any involvement of anybody in the crypto-community educates and helps build the ecosystem.

During the 2018 Transform Africa Summit in Kigali. Charles Hoskinson from Cardano,[14] Tunde Ladipo from Stellar,[15] and Makoto Takemiya from Soramitsu[16] each presented solutions and visions from the perspective of two of the top ten Altcoins and the largest open source project, the Hyperledger permissioned Blockchain platform. They showcased how their respective technologies could have positive tangible results for African communities and countries. Obviously, that reflected idealism and they were making meaningful personal and corporate commitments.

The seeds are being sown but not everybody likes the way it happens.

Selling or Sowing?

In response to them, nChain's Craig Wright stated "everyone over there [...] wants to sell you [something]."[17] At least philosophically this can of course also be applied to Craig himself, or why would he otherwise have been there? His point, right or wrong, seemed aimed at short-term vs. long-term financial and commercial interests. When he later compared his personal wealth favorably to his host country's GDP numbers, this did confirm all three points I made above.

It exemplifies the danger of what I generalize here as an "outside-in" approach: One person like Craig Wright could literally buy himself dominating market positions in populous societies, for just a fraction of his personal wealth. That definitely looks like familiar historic patterns, in an eerie way. And of course, like a no-go approach.

Extreme positions can inject clarity and therefore Craig's bigger point is well worth contemplating: To maximize long-term benefits for Africans and for business activities in Africa, is it the right approach starting with a foreign for-profit organization trying to gain market share by penetrating and dominating small African economies with solutions mostly developed by outsiders?

Doesn't this disproportionally extend financial influence or even sometimes control over large populations? Doesn't it distort markets by injecting priorities from the outside, during that process effectively creating a new intermediary and establishing barriers of entry and limitations for individuals? And finally: Shouldn't one instead focus on "blue ocean" development of African talent, thus leveraging the key resource Africa offers: human capital?!

It is a difference in approach, however, and not of ethics. Platforms compete and one way or the other there are personal and business interests linked to their success or failure. As Nassim Taleb asks for in his newest bestseller – at least all the above have some *Skin in the Game.*[18] This applies to BSIC's Sandra Ro as much as to Charles, Makoto, Tunde, Craig, and their respective ventures and platforms: Cardano, Stellar, Soramitsu, and others that are engaged on the continent, like Bitcoin and Bitcoin Cash.

The priorities seem clear. In the big scheme of things, there arguably are two pressing Blockchain-related tasks to be accomplished in Africa, setting the stage for most other actions. The first is advocating and achieving a basic opening of economies toward crypto and Blockchain technologies. This likely involves facilitating the intermediary-free, high performance, and low-cost flow of money, and government regulation not getting in the way.

The second is generic crypto-education and building an active Blockchain community. Blockchain initiatives need to focus on what drives future wealth. The real treasure and appeal of African societies is not the profit that can be extracted from their relatively limited existing economic activity but tapping the creative and educated minds and brains of Africa's young people.

With such a basis, Africa's heroes can pursue far-reaching Blockchain opportunities. First, though, a few more words about Money *in* Africa.

"I've Got More Money Than Your Country"

Rwanda's GDP is around $10 billion.[19] To eradicate poverty, this number should be closer to $500 billion. Many good things would come from that, but definitely the following statement would then be incorrect. Today it is not.

"I've got more money than your country." This is what nChain's notorious Chief Scientist Craig Wright publicly stated during the Transform Africa Summit on May 10, 2018 in Kigali, Rwanda.[20] While considered an appallingly arrogant statement by many Westerners, the personal feedback I got from several Africans was not at all that negative.

When I later talked in private to other African Blockchain panelists who spoke at the same event, they interpreted Craig's statement as simply stating the unfortunate truth. In their opinion, it gave him even more credibility that he did not mince words when pointing at the fact of Rwanda's and Africa's relative poverty. He then also offered his take on a solution, namely free trade and opening up African economies using a strong and stable common currency and commercial infrastructure that empowers individuals.

Indeed, the most apparent context of Africa's tech communities is the absolute and relative economic weakness of the rather poor African societies they are functioning in. African societies are not only poor in general, but they are also splintered into over 50 separate and often further divided multi-ethnic nations. As a consequence, control of access and cash rule more so than in most rich countries with more economic security and competitive opportunities. This element of the general African economic system also spills over into the Blockchain world.

Most stakeholders in Africa's Blockchain community are the same as those in the rest of the world. Specific to Africa, though, the shortage of hard cash makes for some rather unusual bedfellows. This can most clearly be seen in the role of outsiders and governments.

Africa's small economies and general inefficiencies do not generate enough revenues through the private sector selling products and services. A well-developed venture capital infrastructure supporting private sector innovation and entrepreneurial initiatives is also still largely absent.

Money therefore routinely gets channeled from the outside through supranational institutions like the World Bank and International Monetary Fund (IMF), well-funded non-governmental organizations (NGOs) in the form of big foundations, aid organizations, or even private businesses, and then distributed in some level of coordination with governments. Obviously, this often involves a political process, which then frequently is described as corrupt.

It is different from how most of the Blockchain world on other continents manifests itself. There, and independent on whether the driving motivation is libertarian, anarchist, or profit-seeking, governments, and big businesses are mainly seen as intermediaries to be opposed and replaced by decentralized solutions. Particularly with the rise of initial coin offerings (ICOs) even the access to a new kind of crowd-funded capital has become a core characteristic of Blockchain initiatives, although an increasingly questionable one.

Despite this, humanitarian and often even egalitarian idealism is still widespread in the Blockchain industry. There is a common belief in bottom-up and peer-to-peer (P2P) empowerment via decentralization and trustless consensus mechanisms, and a common urge to bank the unbanked, include the poor into the global economy, and help the disadvantaged.

That should be good news for Africa, but time is of the essence. Rapid population growth and lacking infrastructure pressure the continent's ability to catch up to the industrialized world. I even question whether catching up is the correct aspiration.

The rich world is eyeballing trillions of dollars of investments into *upgrading* infrastructure that Africa just started to build. Using traditional means, this seems like an impossible task. After all, 5G, high-speed Internet, gene sequencing, clean power generation, continuous education, all need huge amounts of money. If that money instead needs to be spent on satisfying basic needs, the numbers just do not add up.

And while some progress has been made in the past 15 years, recent technological trends add new challenges. They close traditional paths for development and Africa risks falling further behind again because its lacking infrastructure makes it impossible to take advantage of emerging technologies.

A look at these developments increases a sense of urgency. It also keeps building the case for different and much more far-reaching Blockchain-based solutions.

Colonialism was motivated by the search for raw materials, from diamonds to agricultural products like coffee or bananas. Nowadays this would also include cobalt, crude oil, and natural gas. However, making agriculture or mining the linchpin of a development strategy does not make sense. These industries are not just subject to significant swings in commodity prices. They are also not even remotely scalable enough for the aggressive advances Africa must make. Using Blockchain technologies to streamline supply chains of coffee or mining products is likely to create benefits but is in essence only incremental progress.

So, what are Africa's options?

Chapter 3

An "African Tiger" – Does Such a Beast Exist?

SINCE WORLD WAR II, many societies have managed rapid development from a low level, usually with a lasting impact on the rest of the world. First was Japan in the 1950s, joined since the 1960s by the Asian Tigers of South Korea, Taiwan, Hong Kong, and Singapore. Starting in the 1980s, particularly the 1990s and 2000s saw China ascending. It now seems India's turn.[21] And a number of indicators show that Africa may be next,[22 23] or that the process may already have started.

So, is the sequence as follows?

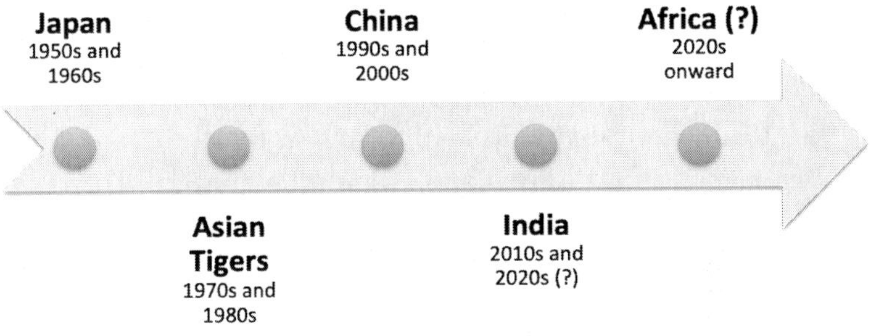

Figure 1.
Rapidly Developing Societies

Of course, the story is not quite as simple as depicted here. Africa is a diverse continent, with lots of unique challenges. Key questions remain. Does Africa have a realistic long-term strategy in line with general economic and technological trends? Does the traditional economic growth model starting with low-cost labor-intensive manufacturing still apply? Will just some African countries develop rapidly, or can the whole continent be transformed?

Another China? India? Or Following Your Own Path?

The ongoing rapid development of India may give some indication. India and Africa are very similar in many respects, in population (1.2 billion Africans vs. 1.3 billion Indians),[24] the number of young people (800+ million Africans vs. 700+ million Indians), diversity of cultures and languages (according to Wikipedia ca. 29 official and ca. 1,250 to 2,100 native African languages vs. 22 official and 1,652 native Indian languages),[25] [26] and overall size of the economy (in 2016 ca. $2.33 billion for Africa vs. ca. $2.69 for India).[27]

There are three main differences, though:

- **Geography.** Africa is much bigger than the Indian subcontinent, more than nine times the size of India. Only Asia is larger. Geography alone makes it much more challenging to treat Africa as one entity for business or economic purposes, and even considering its underlying technological infrastructure.
- **Political and Economic Coherence.** Africa is not politically and economically unified. On the other side, India has one currency, one military, and one political, legal, and economic system. It also has one common governing body, although with large aspects of decentralization into its states.
- **Governance and Political Stability.** Africa is politically less democratic, more diverse, and has less stable living conditions. Only a few countries have established deep-rooted democratic and legal standards, traditions and practices.

Unlike India, Africa is overall less peaceful.

Taken together, the above points may give doubt to whether a continent-wide transformation can happen. Some severe and profound changes are needed, to the underlying political, economic, financial, legal, and technological infrastructure. History doesn't help either. There is no precedent for treating Africa politically or economically as a unified entity.

Although most African societies have many commonalities in the living conditions of their people, the continent's fragmentation continues to be a logistical challenge for any commercial activity. It also makes it very difficult to scale out. This needs to be overcome, and I will later show how it can be achieved, not the least using creative Blockchain-enabled investment and business models.

Low-Cost Labor Manufacturing No More

Before we get there, the big question will be how far a low-cost labor manufacturing approach can go in Africa, and how successful it can be. China, and again India, may be able to give some clues. By using its competitive advantage of low costs for (mostly physical) labor China became the global "extended workbench" and the world's manufacturing hub.[28] It lifted hundreds of millions of people from abject poverty. And still, there are debates whether the country can escape the middle-income trap, which is the term to describe nations whose economy grew rapidly until they then got stuck before their population became rich.

Starting with the automobile industry, a "Make in India" campaign now tries to achieve rapid growth by replacing China as the manufacturing powerhouse of the world.[29] The Indian strategy is also driven by leveraging low labor costs. This time, the viability of the approach is in question.[30] Manufacturing trends are again starting to shift.[31]

Automation, robotics, 3-D printing, and personalized production are increasing the digital contribution of manufacturing and changing where things are produced, and where and how people create value.[32]

New technologies reduce the portion of physical labor cost, subsequently affecting employment. In simplistic terms, if 50% of your costs are related to physical labor it makes much more sense to shift manufacturing to a place where one can save 80% of that portion. One then saves about 40% of the overall costs. However, if only 10% is physical labor-related, then the savings are much lower, just 8%. In the latter case, other factors may be more decisive than costs, like quality, shipment costs, branding, timing or risk.

> "By 2020, 50% of the G2000 will see the majority of their business depend on their ability to create digitally-enhanced products, services, and experiences. [...]
> By 2020, 75% of F500 companies will be suppliers of digital services through Industry Collaborative Clouds."[33]

Perhaps India still manages to reach its goal to soon employ 100 million people in manufacturing.[34] But there are already indications that it may not. For Africa, though, the chances shrink yet further to again repeat this same strategy and succeed. The continent is even further behind in key aspects like supply chain integration, training, capitalization, skills, and general infrastructure. It, therefore, remains doubtful whether traditional manufacturing can be Africa's basis for rapid and sustainable economic development.

This is not the end of the story, of course. A look at the future can give us some pointers for how to solve the conundrum.

"It's the Brains, Stupid:" High-Tech is the Answer

Technology-based services require no more than educated brains, an Internet connection, and an effective platform leveraging these brains. As far-fetched as it may sound at a first glance, they could be the most natural fit for Africa's strength. This would also make them the best bet for African medium and long-term economic success.

On a basic level, technology can facilitate the creation of value in energy, health care, and many other industries, or by providing general services. More important, though, it also can create exports value via processes similar to today's outsourcing. African designers, developers, or analysts can inject themselves into increasingly digitized global manufacturing and service supply chains.[35] They can create new products and solutions. Neither physical goods nor locality matter as much as before, and digitization and personalization are becoming the main drivers of value.

Africa has more young brains than any country. Education, high-speed low-latency Internet access, and an effective commercial platform can tap deeply into those. More about that later when I describe how new economic models can entice investments into these core infrastructure components. African entrepreneurs are then enabled and can figure out a way to take advantage of opportunities via future-oriented digitized products and solutions.

None of this can be achieved top-down, though, because it is part of a larger global and mainly decentralized network of economic activities. It requires the opening up of a free and efficient flow of data, whether general information, electronic money, or supply-chain related business transactions. Focusing on flexibly integrated services delivered by African brains into global digitized supply chains, this becomes an Internet of Brains (IoB), related and equivalent to but distinct from the Internet of Things (IoT). A highly responsive rules-based network of people, of human brains, can deliver high-quality services in a transparent manner.

A Matter of Trust

Decentralized crypto-solutions leveraging Blockchain have to be core components of any such strategy. They unleash the power of otherwise disjointed groups and individuals, without needing to trust centralized hierarchies. Crypto-technologies can seamlessly integrate high-value-creating individuals into a global network delivering digital products and services, or digitized components surrounding physical products.

The distinct appeal of Blockchain technologies is their focus not on hierarchy and rigid linear structures, but on the empowerment of individuals in a larger and rules-based community. Hierarchies and rigidity benefit those on top. This effectively limits the scope of what is possible to the leader's level of ingenuity and ability to execute.

On the opposite side of the spectrum, an organically growing community trusting the power of decentralized individuals and groups does not inherently contain such artificial limitations. Decentralized and organic communities do neither do as told, nor copy, but adapt to their specific circumstances. They maximize value by preparing for the future. No trust in centralized leadership is needed.

This is then how Blockchain and crypto-technologies can shape the new Africa in a most dramatic fashion. They can trigger and enable the development of new economic and governance models that work on a continent that otherwise is not unified. One of the main components of this is a free flow of money, likely in the form of cryptocurrencies. Another one is the construction of future-oriented communication and computing infrastructure.

The same technologies can also address real-life and often-existential challenges and dramatic opportunities for Africans. As happened with the leapfrogging of telephone landlines and the development of mobile money, Africa is unlikely to simply emulate models of the industrialized world. With global value generation increasingly shifting to services and digitized products, even the value of traditional "industrialization" as such is questionable. Developing centralized energy grids would be as backward oriented as using landlines. So, would be spreading the use of credit cards for cross-border transactions. Or the expansion of the wide-spread 1880s German-style health insurance schemes that dominate North American and European societies, with their silos and intermediary-focused inefficiencies.

Blockchain's disruptive potential goes further than that. It can replace governmental record keeping on traditional cloud or otherwise centralized databases that require trust in people and the resilience of system. Blockchains lower risk by eliminating single points of failure and attack. The decentralized managing of agricultural supply chains can empower farmers and cut out middlemen and non-value-adding intermediaries. They can make logistical chains more reliable and efficient. Crypto-coins can be the tools to establish markets for sharing tractors, energy, or storage space. And so much more.

Nothing, though, can even be remotely as powerful as sailing into the "blue ocean" and creating profoundly new and scalable revenue opportunities leveraging emerging high-value technologies.[36] That is what satellites, 5G, genomics, Robotics, IoT, artificial intelligence (AI), augmented and extended reality (AR and XR), renewable energy, and new batteries are all about. Enabled and probably even powered by Blockchains.

Chapter 4

Filling a Near-Empty Glass

IN HONG KONG, at a conference about emerging technologies, I met the "father of Africa's Internet."[37] A native of Ghana and educated in the United States, Nii Quaynor pioneered many aspects of Africa's Internet development and expansion. When he presented at the MIT Emtech Hong Kong conference on June 7, 2017, for the first time I realized *how* critical of a condition data communication infrastructure really is. The brains' promises are unfulfilled unless they can access and interact with the world! I felt like a little kid, surprised to learn that electricity does not "just" come out of the outlet in the wall, but that there actually is a whole industry behind it. I "knew" that, of course, but I never had really "understood" it.

After his presentation, Nii and I spoke for a few minutes, and his unassuming and amiable demeanor showed something else. A situation that looks near hopeless is only so if one accepts that it is. Heroes do something about it. Like Nii, African heroes then implement an ambitious vision and bring the Internet to Africa, in the process changing a whole continent. No surprise then that in 2013, Nii Quaynor was inducted into the "Internet Hall of Fame."[38]

Similar to the Africa of early 1990s, almost totally without a data connection to the outside world, other aspects look near hopeless also. Many people see the rapidly growing populations of Africa and conclude that there are too many mouths to feed.[39] This glass-half-empty, or "glass-near-empty," look becomes self-perpetuating and leads to a hopeless dead-end. In reality, though, we can also fill the glass up. One just has to look at different organs than stomachs: There are so many young brains to tap into! Focusing on these brains enables us to look at the *opportunity* rather than the problem.

If we see and acknowledge the dramatic potential residing in the most important raw material of the 21st century, human brains, then we no longer need to *provide* for Africa. We *invest*, for our benefit and for the benefit of those that we invest in and the societies they live in. Then we can see how the world benefits from contributions that African brains can make. Everybody wins. We get immediate benefits from investing in African brains, and "we" means Africans and non-Africans alike.

Investments fill half-empty glasses, and even near-empty ones.

Blockchain is a natural tool to turn the old ways of thinking on their head and empower new ones. They can motivate investments by tokenizing assets and their use, mathematically enforced and outside traditional hierarchies and political control. No longer do large portions of African societies need to rely on being fed by others. Instead, smart Blockchains reflect the belief in the power of the individual woman and man, in the capabilities of each and every human brain. Blockchains provide avenues for entrepreneurial activities and an individual's pursuit of happiness, without being limited by gatekeepers.

This is why, whenever I talk to a new audience about Blockchain, crypto, or the decentralized world, I usually start by asking three questions: Who likes corruption? Who likes being constrained? Who likes being controlled?

Then I stress that removing such limitations is what Blockchain is about, or at least can be about.

Crypto-technologies like Blockchain can make it very challenging to "game the system" and gain advantages by being a middleman or gatekeeper. This is how they *reduce corruption*.

Blockchains can redefine organizational structures so that they become more organic, self-organizing, and dynamic. They then better reflect the natural complexity of systems. This liberates them from the traditional rigid, artificial, and inefficient linearity and hierarchies that limit people's success in general, or their individual pursuits. It is how Blockchains *eliminate constraints.*

By having nobody specifically in charge of these self-organizing communities, crypto-tech promotes individual economic freedom, and thus also personal freedom. People no longer must be subjected to control by others, particularly not when living their daily lives and pursuing their own dreams and ambitions. This is how Blockchain technologies *liberate people from being controlled.*

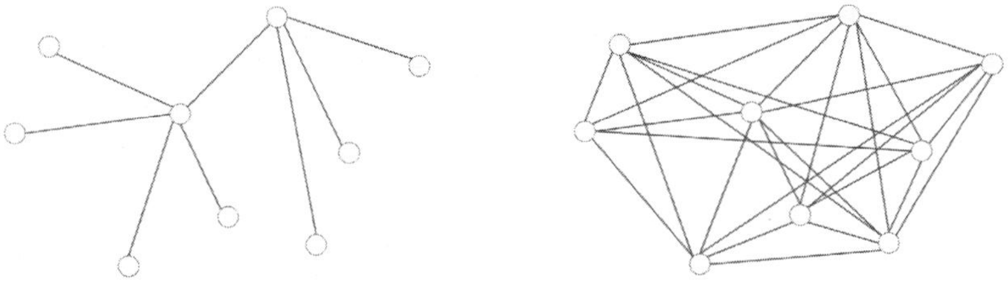

Figure 2.
Networks with Intermediaries (left)
- and a Less Centralized and Therefore Empowering (right)

Blockchains create an opportunity for new deals, for new structures and economic models in many if not most aspects of societies, including political and business organizations. They have the potential to empower individuals and groups in a bottom-up manner. With that process, they *create opportunities for people who previously were limited or excluded.*

Technology can become the natural enabler, and Blockchain is a prime example of that. Its mathematically-governed, economically-motivated, and decentralized infrastructure means that in principle it is almost instantaneously scalable in an almost unlimited manner. My main point is that using Blockchain technologies to facilitate the creation of a pan-African technology-based economy is the most effective, most scalable, and most organic and therefore lasting approach to interact with and succeed in the African market.

How to Unleash 800 Million Brains?

Africa has many heroes. Too many, though, are leaving the continent because they perceive a lack of opportunity. Their children find success and inspire the world, by winning World Cups for their new home country France and commanding the world's most powerful economy and military as President of the United States of America.

Nothing is wrong with emigrating. I did it myself. A lot is wrong, though, if the reasons for leaving the country one grew up in are undignified living conditions and a profound lack of opportunity.[40] Nothing can therefore be more important than enabling African heroes, actual ones and prospective ones, to *thrive, not survive* – in Africa.

We all know that Africa's 800 million young brains are being used every day. Children, women, and men use them to navigate the myriad of hurdles, pressures, and limitations in their daily lives. They hustle and create income out of nearly nothing, whether washing cars or peddling fruits. Many already sit in innovation hubs behind their laptops, using Internet access and learning about technology, from coding to entrepreneurship.

But they are constrained and their opportunities are limited.

Removing these restrictions and limitations is not a miraculous task requiring a monumental governmental effort. Most of what it takes is enabling the free flow of data and commerce, from contracts to money, and an open mind about new economic models to fund future-oriented education and infrastructure in innovative ways.

Old models had only limited effects, often none. Distribution of wealth initiated by a "socialist" elitist government quickly runs out of steam, or what Margaret Thatcher called "other people's money."[41] They also inherently limit freedom of individuals since they shift authority to politically controlled and usually centralized bureaucracies.

Freewheeling "capitalist" approaches based on individual property rights break down such barriers of regulation and interference in individual's lives. By doing so, however, they tend to strengthen existing power structures, thus establishing intermediaries, introducing monopolistic inefficiencies, and perpetuating inequalities, particularly when they started unevenly in the first place.

Humans are by their very nature competitive, but also communal. Since people can relate to both sides of this coin, many societies yo-yo around between extremes.

There is an alternative, able to better reconcile these perceived contradictions of individualism versus communalism. New ways of thinking and modeling can take emerging technologies, apply different economic motivations, and use tokens to enable previously unachievable solutions for whole societies. When such new organizational principles consider the general interests of human communities and societies, they can supersede century-old barriers. Private initiatives can then flourish and unleashed individual brains can do the rest to improve everybody's lives.

Tokens (or coins)[42] have this power to replace traditional concepts of ownership and control. They enable people to act in pursuit of their self-interests, driving the token-value up, and thus automatically and indirectly benefit everybody else in the community.

Such decentralized and distributed systems are then linked by common interests. They are also stable and resilient because they are not controlled by anybody for any specific individualized purpose. To take advantage of this opportunity, Africans benefit if they work together. A common continent-wide ecosystem and commercial platform can smooth out imbalances and eliminate dependencies caused by the splintering of an already relatively small economy into over 50 nations and hundreds of economic centers.

All this is possible with much less investment than people think. It does not require complicated politics and governmental leadership either but can be complementary to programs that too often rely on limited governmental budgets. A big part of health care does not have to only be about having doctors, equipment, clinics and hospitals to treat sick people. It can also be about finding and funding preventative solutions that preemptively compensate for genetic defects.

Energy does not require large dams and multi-billion-dollar infrastructure projects. Put solar panels on each roof that today already has a satellite dish, add a battery, find a new economic way to fund it, and most energy issues are resolved. Micro-grids, nano-grids, and household-based energy independence work in decentralized fashion, but they can scale out organically, and powerfully so. Education does not need to be, even cannot be, mainly provided by governments, particularly considering that we are entering a time of continuous learning in an ever more rapidly changing world.

Opportunities for significant improvements are everywhere. Africans can use genomics to analyze soil and better provide food. More efficient means to do agriculture can save efforts, increase profits, and reduce prices. Even the turnaround time for creating new genetic variations of hardened and more bountiful crops can be cut short. Communications can be provided via micro-satellites, 5G relays, and new access and payment mechanisms for Internet access.

The real power of Blockchains focuses on such matters instead of incremental solutions that are not transformative.

My main argument here is that the focus is important. If we concentrate on the problem, then we miss out on the opportunity. If we focus on "catching up" we cannot get ready for the future. The real power of Blockchain requires going beyond incremental solutions that are not transformative. But if the vision for the future is to prepare for the almost completely digitized economy of the mid-21st century, then trying to catch up to 19th or 20th-century solutions is not the right approach.

Rather, everything needs to facilitate the tapping into the brains and minds and unleashing their power. If we could train all these young brains today for three to five years, or even substantially less, they would be on par with comparable brains in the rich world. Human knowledge and technological advancements move so fast. There is no need to catch up over decades. It is called *leapfrogging* and Africa has done it before with acquiring mobile phones instead of landlines, and with mobile payments instead of traditional banking.

Chapter 5

The Trillion-Dollar Argument

IN HIS OWN WORDS, Michael Kimani "can't wait to change the world."[43] He does it by "making" money – literally helping people to build, improve, and use cryptocurrencies. In Nairobi, where I first met him as a co-panelist at iHub, he has been working with cryptocurrency and Blockchain initiatives like Ethereum, Bitpesa, and EOS. It was not always like that, since he got involved with Blockchain just about a handful of years ago. By now, as the chairman of the Blockchain Association of Kenya, he also is a co-founder of Chamapesa, a Blockchain identity startup. He spoke at our *TheWorkshop* Blockchain event during TAS2018, and one believes him that he indeed will change the world.

Although Maximillien Bingabo is new to the Blockchain world, he also presented at the same event. To his young African audience, he explained how he, a developer with Rwanda's WiredIn, picked up Ethereum development skills, and within a couple of months was able to contribute to a Blockchain project.

When he heard that I would be in town, Markos Lemma, the high-energy CEO of Addis Ababa innovation hub and incubator iceaddis, organized the first Blockchain meetup in Ethiopia. Within a couple of days this self-acclaimed "techie, activist" brought together a sizeable group of attendees at iceaddis, in December 2017.

This is where I then also met Natnael Berhane from Mauritius. He approached me after my talk. Still a student, his vision is to start a regional exchange platform.

One step leads to another. Africa's vibrant tech communities are already informally connected. For them, it usually is only *one* or two and not the six degrees of separation that Duncan Watts talks about in his ground-breaking book.[44] The above are just four of the brains and heroes that I keep talking about. They have in common that they are ready, able, and willing to learn and contribute and make a dent in their world. And so they do.

If one looks for customers, Africa is usually not a market worth getting into. People there simply don't have enough money. If an outsider tries to make money in Africa, they can either live with a small piece of a small pie, or they can leverage personal connections and intermediaries to get a bigger piece. Either way, the pie will remain small.

For many young African entrepreneurs, this may be okay. A small piece of a small pie may still make for a very good living in an African country, relatively speaking. Solutions to real problems may also have a big impact in a small space, and possibly even beyond. However, today there are inherent limitations and it is next to impossible to scale this out without changing the rules of the game. African entrepreneurship needs to be stimulated much more strongly.

Blockchains do that by introducing new models for economics and governance. Their impact can be nothing less than dramatic. When investments can be effectively governed outside the influence of intermediaries and gatekeepers, then business cases just need to show satisfactory returns. And when these activities involve the selling of high-end digitized services and solutions, those returns can be breathtaking. Even more so when put in the context of "traditionally" low-income countries.

Michael, Maximillien, Markos, and Natnael can make it happen (as can the before-mentioned Norbert, Njoki, Amadou, Alain, Dawit, and many more). What they have to offer is something extremely valuable to the world. It is their brainpower, entrepreneurialism, creativity, and hard work. They are by far not alone. After our workshop in Kigali, dozens of students, professors, young entrepreneurs, and tech enthusiasts kept hanging around to talk to us.

They are on par with the best anywhere in the world.

Why not cause 15% to 20% growth of an economy by "betting" on the output and economic value that these African young brains can generate? In ten years, this would mean an economy would have grown to be four to six times as large as it is. Almost all of this could be created by focusing on the emerging digitized supply chains.[45] Why again would this be impossible? Because of engineering challenges? Not so. Because of inflation? Cryptocurrencies have solutions for that also.

Here is a simple calculation, using Rwanda as an example. Let's assume its population to be 10 million people and its GDP to be $10 billion in US dollars, which equals a GDP per head of $1,000. In reality, the country's population is slightly higher and its GDP is lower. Over the course of ten years, if we would have just 1% of the population, that is about 100,000 young minds, contribute 2,000 hours a year to the digital economy and generate value and revenues at $50/hour – Rwanda's GDP would have doubled. Just this capital transfer from abroad equals an *additional* 7% of the annual growth of the GDP for ten consecutive years, not even accounting for the trigger and trickle-down effects!

To reach a comparable result continent-wide, Africa would need to reach a total employment of 10 million people that are fully integrated into the global digitized supply chains. The potential economic impact of that alone is no less than ***$1 trillion.*** This is just the direct impact, not considering what such an influx of capital from the outside into African economies could do. And it is still far less than the 100 million that India wanted to employ in the manufacturing sector by 2022, following the old low-cost labor model.[46]

That part of the traditional equation remains the same when compared to the old low-cost-labor growth models: it creates a capital surplus. And this is exactly what low-income countries need. *Money.*

To reach a six-fold increase of the GDP in a decade, this would require injecting about 5% of the population into the global digitized supply chains.[47] All that is needed is Internet and computing access and education – and a commercial and financial infrastructure free from interference.

Do this for the whole continent, and a combined economic power the size of today's China has been created! All in one decade and all by unleashing the potential of *some* of the brains and enabling them to participate in the big technological and economic trends of the third decade of the 21st century.

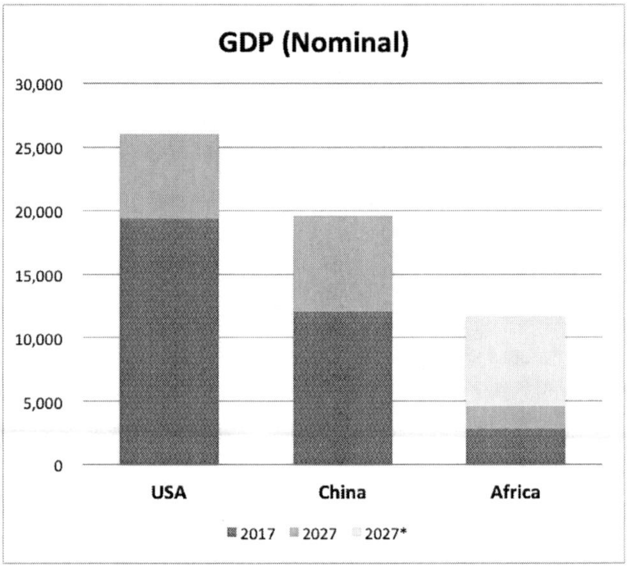

Figure 3.

GDP Numbers for the U.S., China, and Africa

2027 projections are based on 3% annual growth in the U.S., and 5% annual growth in China and Africa. 2027* includes 5% of the population generating "blue ocean" income of $100k per year (income is overall revenue based on value generation, not necessarily personal income).

This may sound too outlandish, but it simply exemplifies the outer limits of what is possible. Obviously, success depends on many details and above all on effective execution. It clearly shows, though, where real transformation can come from, and what leapfrogging and transcending of large societies look like. It is realistic if Africa manages to create Blockchain-enabled economic models that use its competitive advantage of young and capable brains to leverage emerging technologies.

Any higher efficiencies and gradual improvements are positive and contribute to decent growth and advancements. Still, a future-oriented transformative strategy involving technology and Blockchain investments and projects in Africa *cannot* focus on slowly catching up or be about selling solutions *to* African nations. It should *not* be about convincing some foundation, government, development organization, or supra-national institution to give money so that incremental improvements can be achieved or tried out.

African heroes can do whatever it takes to reach or get close to the big vision, powered by an unleashed entrepreneurial spirit and capabilities of African minds and brains. Aligning their motivations via Blockchains and crypto-economic motivations guided by tokens, Africans together with global entrepreneurs and investors can leverage advances in computing, communication, manufacturing, genomics, energy, and space technologies. Toward the end of the book, I will outline how to develop such transforming and transcending solutions in a few courageous but also relatively straightforward steps, and what particular role Blockchain plays in it.

Current power structures are to be ignored, turned upside down, or complemented by introducing more distributed solutions that leverage Africa's greatest assets in line with long-term technology trends. Everything else is just tinkering on the edges.

First, though, we must dive more into details of the specific context and characteristics of the African continent. These define the context that our African Heroes and their vision for Blockchain have to operate in, and the parameters for additional challenges to overcome.

This is what the second act of this book will do.

Act II

Prepare to Launch: Africa's Path to Wealth

* * *

Chapter 6

Blockchain Promises and Premises

AMERICAN FOOTBALL has the idea of a "Hail Mary" pass. This is the play when all seems lost at the very end of the game, and a quarterback throws a pass as far as he can toward the opposing team's end zone, hoping for somebody to catch the ball and win against all odds. Hail Mary passes are grandiose gestures. They have a "catch," though. Usually, they are followed "by a prayer," because they almost never succeed.

How is that different from claiming to know a way of achieving in a mere decade or two what eluded African leaders and their domestic and foreign supporters for over half a century? A cool vision and some back-of-the-envelope calculations do not make a plan. They are not even a strategy, and maybe just another half-baked idea from an outsider.

A vision needs a plan that realistically considers practical circumstances, constraints, and obstacles, and devises ways of tackling and overcoming actual barriers. In the following chapters, I will outline key elements of such a plan. The big vision is achievable, as long as the plan focuses on a blue ocean approach that could, in theory, be executed anywhere. We just need to apply it to the special context of Africa so that the continent can join in from the start and need not play catch up again.

Importantly, the plan is a business plan and not a political or social one.

None less than the late founder and much-revered leader of Singapore, Lee Kuan Yew, helps us figure out the difference between plans that work and those that do not. In four words, it comes down to *pragmatism*, *results*, *entrepreneurship*, and *education*. More literally, at the African Leadership Forum on November 8, 1993, he shared his wisdom for African development, including the following four of seven principles.[48] He said, "let foreigners and local entrepreneurs get on with their business," "pragmatism not dogma in economics," "go for results, not political correctness," and "educate everyone."

This is a good summary of the above outlined Blockchain-enabled vision. It is about ***creating a way for local and foreign entrepreneurs to go beyond classical economic dogmas and instead use educated people to deliver tangible results for Africa***. Blockchain technologies do not specifically prescribe to anybody how to solve complex problems. But they can establish platforms that empower large networks of smart people to achieve a common goal. This unleashes the people's entrepreneurial spirit to deliver high-value solutions that the world will pay good money for.

The key power of Blockchain-based economic models is their focus on *decentralization* and *mathematics*. They appeal to people's self-interest, but in a way that benefits a larger purpose. As such they are a-political and non-doctrinaire, unlike many other traditional prescriptions designed, delivered, and controlled by elites and governments.

Beyond Math: Obstacles and Victories

So, a glorious Blockchain-induced future for Africa may be just around the corner. Or not. There always were bright spots in African development after colonialism, but somehow little stuck. Africa is not alone in this. Latin American giant Brazil has often been said to be the land of the future, and it always will be. It always will be because it just seems unable to ever arrive there.

The lofty goal to radically transform Africa by empowering its many young brains with emerging technologies will remain an unattainable dream unless it manages to align perception and reality. Mathematics alone cannot transform a country, let alone a continent.

The heroes building the Africa of the future must consider and prevail over a number of real issues first. Some seem insurmountable at first. Luckily, none are. In the big picture, there are five main obstacles. All are universal challenges, but each has a unique African dimension. They are:

1. **Confidence.** The lack of confidence in Africa, by Africans and foreigners.
2. **Infrastructure.** Africa's infrastructure being insufficient for ambitious goals.
3. **Fragmentation.** Africa being geographically, economically, and politically fractured.
4. **Governments.** The often-constricting role of governments in Africa.
5. **Funding.** The absence of an adequate investment infrastructure in Africa.

By addressing these challenges and solving them, the general outline and shape of the overall solution will become ever more apparent. Then we "just" need to add Blockchain and we are ready to go.

At the very beginning, one needs to figure out whether our prospective heroes of Africa's future are even ready. After way too many frustrations, do Africans still believe in themselves? And since the vision requires partnerships with the rest of the world, how do foreigners see the continent? Blockchain is about self-organizing networks of people whose motivations are aligned. How can such a community be created?

The nearly all-present lack of infrastructure is the second obstacle, and a massive one. Africa is a continent three times the geographic size of the United States (U.S.) and almost seven times the size of the European Union (EU). Building comparable transportation systems would likely cost a multiple of the investments made in these rich continents over many decades. Such investment would need to be funded by an economy only about $1/8^{th}$ the size of either the U.S. or EU, although most of the continent's income today has to pay for basic necessities of life, like food, clothes, and shelter.

Considering these challenges, how can Africa's Heroes leapfrog and focus on the future if they still do not have nearly enough and adequate roads, medical care, education, energy, commercial and financial infrastructure? This is the poorest continent, and by a margin. Good faith cannot miraculously create infrastructure where none exists. How can one prioritize and strategize? Can transformative capabilities of Blockchain help?

Even if it were possible to overcome the problem of missing infrastructure, the continent still will be fractured into a large number of political entities, hundreds of ethnicities, at least as many languages, remain largely dependent on former colonial powers (mainly two), and most recently also have become a playground of geopolitical competition between the U.S., China, and Islamic movements. A consistent and coordinated approach seems out of the question.

For argument's sake, let us assume there was an answer to the hurdles of missing infrastructure and fragmentation, overcoming such adversity. Then one still would need to deal with over 50 governments, many of which are perceived as autocratic, self-centered, corrupt, and inefficient.

It seems like a challenge without a realistic chance of success. Too many have tried, but who has succeeded? Small victories are celebrated, but they are not meaningful enough in the big picture.

And there the circle closes. It just plainly seems hopeless, and one can get discouraged and just give up before even starting. Traditional approaches can obviously not work. A feasible plan needs to think outside the box and make business sense.

In devising such a plan, we realize that all the above challenges leave out the most important ingredient for future economic success, a source of massive value-generation and financial income, and one that Africa leads the world with.

What we need is to creatively bootstrap the actual generation of such value and creation of such income, and to solve the above challenges by focusing on nothing else.

We need to talk about brains.

The Real Source of African Power

In almost any respect African economies and businesses do not matter on a global scale. The continent is fragmented into 55 nations. Despite all the natural resources they contribute to the global economy, their combined nominal GDP just about reaches that of France or California.[49] [50] With no more than 3% of the global output,[51] this is way below Africa's 15% share of the world population.[52]

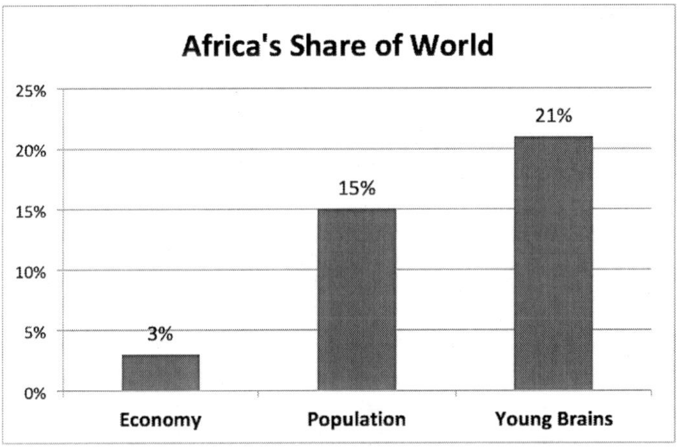

Figure 4.
Africa's Share of the World's Economy, Population, and Young People Below 30

However, over 21% of the world's youth is African.[53] The total number of Africans below the age of 30 is about 800 million,[54] bigger than the comparable number of any individual nation on Earth, including China and India, and far ahead of Europe or North America. *This* matters: due to the continuous shift toward information and data-driven economies and societies, hardly anything counts more than brain power.

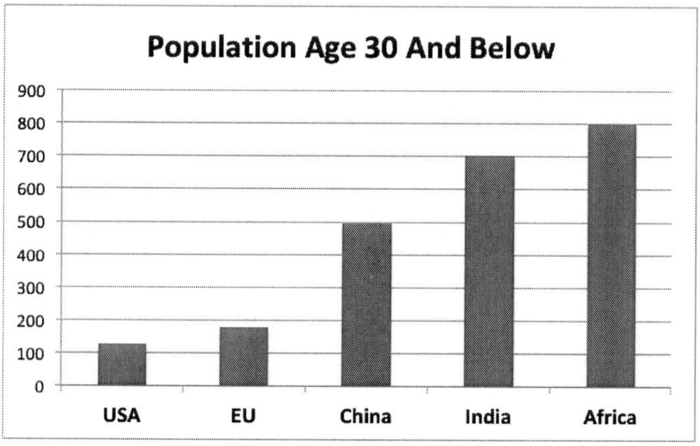

Figure 5.
Population Below Age 30 for World Powers, and Africa

Africa's strength and wealth do not lie in its forests, plains, or in the ground. Its future will be shaped by the brains and minds of its people. Successfully unleashing these can genuinely, successfully, and rapidly transform the continent and positively affect all of humanity. This is not optional since there is no realistic alternative. Failure would likely be disastrous for African economies and societies. Other societies will shift toward a digitized economy anyway, and if Africa does not join them right away it will fall behind again. Technological and economic advances in the rest of the world will then simply increase the gap to the "average" African.

To bring about the necessary unleashing of African brains, restrictions caused by political and economic borders must fall. Outdated structures and dependencies imposed by special interests of Africans and outsiders have to be eliminated, or simply complemented, ignored, and transcended. Individuals are to be empowered, so that they can develop new solutions to energy, agriculture, finance, and health care. These solutions must organically grow inside out and bottom-up. They need to leverage such technologies and new economic and governance models based on the one raw material that matters more to humanity than anything else: the grey matter of human brains.[55]

The key to Africa's future is therefore not major capital investments or ground-moving infrastructure projects but above all, it is developing and applying new skills and ways to "do things." Africa's young want and need to learn them, and then be able to use them, applying their creativity. This statement is supported by virtually every interaction I had with African technology enthusiasts, entrepreneurs, investors, and business leaders.

Bad Technology – Good Technology

There is the before mentioned big *but*, though. Today's changing technology landscape will make it unlikely that traditional models of development can work. What succeeded for a number of Asian countries particularly during the second half of the 20th century is unlikely to still be effective for the Africa of the third and fourth decade of the 21st century. Asia's efforts were centered on attracting low-wage and labor-intensive manufacturing and carving out niches in the supply chains of a globalized world. With its "Make in India" initiative, the next billion-people Asian giant already is tackling a new variation of that strategy.[56] However, automation reduces the value of low-skilled labor-centric approaches,[57] while complex supply chains are breaking up and are being digitized.[58]

These developments cannot be reversed and efforts trying to counter them are unlikely to succeed. Manufacturing output increases while using less labor.[59] Africans would need to compete with India for a pie whose size is shrinking, relatively and maybe even absolutely.

A focus on traditional manufacturing can hardly succeed. Economic development cannot be achieved in a zero-sum game with reduced margins. Instead, it is more rewarding to immediately shift toward digital, personalized, and thus more value-generating skills.

When and as Africa embraces future-oriented technological trends and developments, technology becomes the great enabler and equalizer, and none more so than decentralized but integrated Blockchain-enabled or Blockchain-driven solutions leveraging AI, IoT or the Tactile Internet, AI, 5G communication, and biotech.

The effects will not just benefit Africans, but the contribution of close to a billion young brains and minds will advance humanity in general. Nobody should think they can ignore this. And still one would be hard pressed to find many people that see it that way, Africans and foreigners alike.

Even some of the smartest people often stay stuck in pessimism, as I keep experiencing.

One of my most rewarding activities is learning about a society and culture by talking to people I meet – just about anywhere. This enables me to collect more unscientific but still often insightful anecdotal evidence. Sometimes it does not take a degree to relate, know, or be insightful. Have you talked to a cab driver recently?

Or maybe a cab driver with a degree

Chapter 7

Confidence. Is Africa Ready?

I CALL IT the *UBER ACADEMY.*

On a recent trip, I arrived in Nairobi and took an Uber from Jomo Kenyatta International Airport to the city. As often before, I asked my driver named Joab how things are going in his country. He responded that the rich are getting richer and the poor are getting poorer, and that the main reasons are corruption, special interests, and too much politics instead of a focus on the real problems.

This is not an untypical response in Africa. When I dug deeper, though, a different picture emerged.

Joab is a college-educated engineer. He has a full-time job and uses his personally owned car to supplement his income. Despite having three young kids and having to support his parents, he is able to save money. His kids are likely to get a good education.

I summarized what I heard: “So you are an educated man who is able to not only support your family but save money. You are better off than your parents whom you support. Your kids will even be better positioned since they will get a superior education and will not have to support you when you get older.” Joab confirmed that my statement was correct, to which I replied, “so your generation is better off than your parents’ generation and your kid's generation is better off than your generation. How is life not improving?”

He laughed and acknowledged my point.

Advancing Africa

The first hurdle is in our heads. Do we have the confidence in Africans' ability to deliver? Considering all the negative news, to a large degree based on real-life facts, can and do Africans and foreigners believe that the continent really is ready to directly become an integral part of the global economy's future?

As the above conversation shows, the wants may rise faster than what is being offered (good!). Maybe that sometimes warps perception. But in general, the "stats don't lie," as the saying goes in American sports. Some parts of Africa are getting richer and this directly elevates an ever-broader middle class, and also often the poor. Sometimes this process even gets started by the middle class, like when new kinds of jobs become available in a growing gig-economy, like driving an Uber.

Despite the fracturing into over 50 nations, there are lots of commonalities among African countries. If one takes the broad strokes instead of diving into ethnic and tribal differences, then *diversity* in an African context is not too dissimilar from diversity in a U.S. American perspective. Life in New York City is profoundly different from life in Mobile, Alabama, and even more so from Caldwell, Idaho – although the prevalence of commonalities is obvious. Likewise, European countries differ from each other, but a Finn can feel comfortable in Portugal or Ireland because most core aspects of life are alike.

Although it is not identical, much of the same can be said about Africa. Lives of most Africans are characterized by comparable parameters. Many are negative, like the lack of reliable institutions, underdeveloped infrastructure, a constant need to hustle, distrust in institutions and the motivations of foreigners, even the relative closeness of violence as witnessed by wars, civil wars, or simply the widespread presence of Kalashnikovs and security screening in major hotels and shopping malls. Local currencies, with the notable but fraying exception of the South African Rand, are weak. Euros and Dollars are widely accepted nearly everywhere.

Almost the whole continent uses mobile phones, and increasingly smartphones at that. In many or even most places they replaced bank accounts and have led a global payment revolution. However, the lack of well-funded social systems leads to a focus on cash, in its paper or mobile phone equivalent. In rich societies, not having enough direct income means having to dig into savings, or using credit, often in the form of credit cards. In Africa, it often can mean the difference between eating properly – or not eating at all.

And everybody is hustling and trying to make the best of their lives. Among them are millions of young educated people hungry to learn more about technology. Innovation hubs and incubators and startups seem to exist everywhere. Wherever possible within its limits, Africa is a continent open to technology.

Although many Africans seem to have a hard time acknowledging that their lives are improving, in reality, they can see it themselves. It is just happening way too slowly. Africans dream of a bright future not too different from people anywhere else in the world. Overwhelmed by demands from their daily lives they often have difficulties in seeing it or believing it.

What an irony that of all things it was Hollywood that for many Africans and foreigners alike just recently brought a different version and vision of Africa to the screens around the globe.

Black Panther vs. The Media Trolls

The 2018 movie *Black Panther* was one of the most successful motion pictures of the year, generating revenues of over US$ 1.3 billion and even reaching an all-time Top 10 rank. For the first time, a major movie conveyed a technophile, positive, and optimistic image about Africa, and Africans in general, grossly contradicting the typical picture shown outside the continent.

The power of images is profound. Sometimes a picture needs to be painted, aspirational but achievable, because it triggers and jumpstarts our imagination. Otherwise, if we cannot envision a better future, how can we achieve it? Or as the saying goes: if you don't know where you are going, you will definitely not arrive.

The *Black Panther* is therefore important. Of course, the movie hero's kingdom Wakanda does not reflect Africa of 2018 and is no more real than the worlds of Ragnarok, Guardians of the Galaxy, Ant-Man, or World of Warcraft, for that matter. However, neither is the common storyline about Africa that is told by much of the rich world's media, academics, and politicians.

That image is defined by the media trolls of the world, trying to generate clicks and attract readers by focusing on incomplete and one-sided images. The media in the rich world still commonly portray Africa as a rather primitive war-torn and dangerous place, overpopulated, poverty-stricken, ruled by authoritarians, dependent on external aid, with corrupt elites, abusive big businesses, and control resting with former colonial masters. There is a kernel of truth in all of that – as in some aspects of the *Black Panther*. In reality, though, the picture is much more complex, with lots of reasons for optimism.

Blockchain projects better keep this in mind. Developing in relative or full isolation a solution focusing on "banking the unbanked," this just cannot work. Likewise, the frequently heard African goal of an "African Blockchain," "let's do an ICO," or "we want to create a stable coin" starts with the wrong motivations. Just doing something because it can be done does in itself not create value or solve a problem.

The negative pictures need to be overcome, but we shouldn't rely on Hollywood or the media for that.

What Really is the Difference Between Sh!tholes and Gold Mines?

U.S. Senator Dick Durban popularized the term "sh!tholes" by sharing his recollection of statements made by President Donald Trump during a meeting about U.S. immigration policy. If this word was supposed to describe places where people live in a way that humans should not have to, those can be found anywhere in the world. But yes, such places seem more easily and more frequently to come about on the African continent than in most others.

Many people and organizations in Africa and outside use these images as motivation for trying to better the living conditions of Africans. Although commendable, there is a huge downside to this. The media-driven portrayal of Africa that focuses on poor living conditions and past wrongs perpetuates the problem since it focuses on selective observations of the past and present, and not on the much more relevant opportunities of the future. In a survey we conducted at the 2018 Transform Africa Summit (TAS) in Kigali, Africans seem much more interested in the latter: opportunity and future over problems and the past. More about that later.

It is important to face uncomfortable truths, and then do something about them. More important is that the above does not define "Africa" or "Africans." Most important, though, everybody needs to understand that the intellectual capability of African minds and brains is identical to those in the rich world, and their drive is unsurpassed. Once that truth *really* sinks in, it removes basic doubts and the focus can be on what is possible. Then everybody is ready to build the future, Blockchain-initiated or not.

Arriving in Kigali, Rwanda, a few hours are enough to shatter some of the negative stereotypes about Africa. One encounters clean and in general well-maintained roads, at least the main thoroughfares, a safe environment, friendly and optimistic people, and locals dedicated to providing top services by any standards. This is how Singapore must have felt like in the late 1970s and early 1980s.

Likewise, traffic in Nairobi often is bad, more chaotic and dirty but not really less pleasant than in San Francisco. Within a few weeks, I have been in lengthy traffic jams in both cities. The stress level is equivalent and I wasted about the same time in both places. In Uganda, there are not enough roads, and those that exist are famous for their potholes, but Kampala is also a bustling city ready to do business, full of entrepreneurs including the highest rate of female entrepreneurs anywhere in the world.[60]

Not that there is no poverty in Africa since there may be more here than on any other continent. The negative images, however expressed, are correct in isolation. What they miss, though, is the underlying dynamics and near-dramatic opportunities.

Ignoring the World's Trolls?

The Urban Dictionary defines a troll as "one who posts a deliberately provocative message [...] with the intention of causing maximum disruption and argument."[61] Merriam-Webster defines trolling as "to antagonize (others) online by deliberately posting inflammatory, irrelevant, or offensive comments or other disruptive content."[62] Another term for this is also "clickbait," the practice of being sensationalist and antagonizing enough to generate interest.

With Africa, the world's media, and frequently even academics, volunteers, and politicians, commonly come across as such trolls. Sometimes it is Africans themselves. Being negative can make good short-term business sense since an established image of "poor Africa" is easier to confirm and perpetuate than creating a new one.

But only positive images can inspire and empower. Once they sink in, people want to be part of optimistic developments and join in.[63] FOMO, the "fear of missing out," also is a powerful force.

To be fair, one can also find a more differentiated and balanced view of Africa in the rich world, mentioning hopeful pockets of Africa full of capable people and opportunities. Joab lives a better life than his parents and sees an even better future for his kids. But in theory, he could also sell high-end engineering services on the side and replace and maybe triple his Uber earnings.

There are over 400[64] innovation hubs across the African continent, each with a network from mere dozens to many thousands of skilled Africans. They include some of the brightest people anywhere. It is then basic math. Imagine if they were all connected! They could be more powerful and impactful, thus weaving the platform that launches Africa's future.

Since empowerment requires trusting the capabilities and judgment of individuals, what they need is above all platforms and infrastructure and freedom rather than anything else. It does not take Hollywood to paint optimistic pictures of Africa's future. Reality can do that job. Enough of Africa is not just ambitious but also ready for high-tech jobs, enabled by Blockchain technologies.

One then must wonder why Blockchains can play a role in that process. Overcoming limitations of infrastructure would be a good start. So, let's have a look at that next.

Chapter 8

Infrastructure. Don't Mind the Gap?

AFRICA'S BRAINS ARE STUCK and getting them to move can be expensive.

When you take the bus from Kigali in Rwanda to Kampala in Uganda, you will notice two things. The first one is positive. Besides the muddy border crossing itself, the roads are mostly very nice, and traffic flows well for most of the trip. But then, when you reach the remote outskirts of Kampala, and already can see the city, everything stops dead.

Pedestrians weave their way through traffic, passing cars and buses, even many of the nimble motorcycle *bodas* that usually navigate jams so efficiently. Almost half the duration of the whole trip is spent trying to negotiate the one pathway into Uganda's capital that connects it to the closest million-people metropolis. There is no suitable alternative to this one road into the country's commercial center and main city.

West Africa is not too different. Fly into Lagos and from above you see almost solely dirt roads. Until you reach the city proper, hardly any road seems paved, near sub-Saharan Africa's main commercial center, not counting South Africa.

Such observations may be the most enduring picture documenting poverty for most visitors of Africa. As of 2018, one cannot traverse the continent without encountering dirt roads. Other means of transportation and logistics are also far behind Western standards.

If Africa's heroes want to empower and tap into millions of young and capable African brains, these need to be able to get to work, have power, access the Internet, drink clean water, get medical treatment, and send and receive money. They must be able to take all of this for granted.

Infrastructure? What Infrastructure?

The second big challenge to our vision is a widespread shortage of state-of-the-art infrastructure in Africa. Unfortunately, roads are not the only inadequate aspect. There is a shortage on all levels.

Look at a satellite picture of the world at night, and Africa looks dark except for the Maghreb coastline, Egyptian Nile River, and the major South African cities.[65] Besides that, the second most populous continent looks dark like otherwise only the deserts of Australia and Central Asia, or the rain forests of Central Brazil.

Outside South Africa and the Mediterranean coast, few parts of Africa have meaningful railroad tracks connecting major population centers. Those that exist are not standardized, with a handful of different gauges in use. Most trains are diesel operated. Even today, many railroads are remnants of the colonial era, connecting mining and agricultural areas to ports whose main purpose is shipping these raw materials. The reasons they were built was to effectively control and exploit colonies, not to connect population centers, facilitate commerce, and move goods to or between them.[66]

Fast and reliable Internet access cannot be taken for granted, sometimes even in big cities. Outside the hotspots of hotels and malls, it can be very spotty, and overall slow. Just about 2% of Africans have mobile broadband access, and this already severely hampers productivity today.[67] It will be even more limiting when trying to electronically deliver high-value services in the near future. A major related aspect is the lack of data centers, computing capability, and slow connectivity to content created in other world regions. Often, cloud computing and (not really) free services like WhatsApp are the main options for individuals and businesses.

Leapfrogging Contradictions: Mobile Phones and Finance

Fast and low-latency Internet connections, and easy, quick, and low-cost money transfers are critical conditions for offering high-tech services. Africa must be able to do business with global customers as if they arc next door.

But try sending money from Ethiopia to Kenya, and already then things get complicated and costly. Paying for services from abroad is even worse. It is a shame and somewhat ironic since two of the most visible recent successes of African use of technology involved leapfrogging and directly moving on to the future. Some African societies have successfully done that with mobile phones and banking, and both are connected.

This all started with a simple business case. Instead of laying expensive telephone landlines as the rich world used to do, why not rely on cheaper cellular towers only? Once the resulting explosive use of mobile devices became more widespread than bank accounts, an ingenious idea was implemented to leverage mobile phone accounts as the equivalent of bank accounts. The first simple and widely usable version of mobile money was born.

Considering the prevalence of mobile phones, in a small ironic twist, most phones in Africa still are feature phones, not smartphones.[68] Overall, the use of mobile devices in Africa is still significantly below the average of the rest of the world.[69]

Despite this noticeable digital divide, more notable is that the majority of the African population now can make mobile payments with their phones, is connected, and uses WhatsApp in a unique combination of social and business manners. And they do it like there is no tomorrow.

Since the emergence of mPesa and other mobile money, the average person in East Africa pays with their phone. Many can even get credit with their phones. By combining the leapfrogging of landlines with a clever idea, several Africans countries effectively also have leapfrogged the near-universal need for banks.

While it had very positive effects on some African economies, the picture is more complex. For the person on the street, the effect has mostly been about being reachable via text and social media, making phone calls, and the ease of use for domestic payments. However, in most parts of Africa Internet access is still too spotty and slow, and cross-border payments are complicated, slow, and expensive.

Africans are increasingly connected to each other, but not really to the world in a way that serious commerce could be conducted.

A Role for Blockchain?

Most of the above shortcomings become apparent to anybody traveling in Africa even for just a short while. While in the higher-end areas of major cities most things work at a reasonably high standard, in general roads, Internet, electricity, and even education and health care are not up to the standard most East Asians, Europeans, or North Americans would expect.

Natural questions arise from this. How can Africans then pay for the establishment of much-needed infrastructure? And are there alternatives to the traditional tax and spend models of the industrialized world?

Initial Coin Offerings (ICOs) may not be the universal answer to these questions, but they have received billions of dollars' worth of investments. Often this was for schemes much more questionable than infrastructure development in Africa. So, new token-based economic models may be able to access capital in novel ways.

More elaborate Blockchain-based token models could open the door to decentralized and perhaps superior solutions that bypass and improve on some traditional approaches altogether. For that, they would need to effectively consider the time-value of money and increase the token price by internalizing value into a coin.

That's a mouthful. It means that you may receive a token or coin today, for doing or giving something valuable, and that you can use this token to pay for something more valuable tomorrow. This "something valuable" could be electricity, a technical skill, data transmission, or health data like a genome file.

Instead of copying the rich world's ways of the past, building the infrastructure of the future may be done more successfully using new approaches. For quick success, it may even require them. We can question why to implement a centralized electrical grid when power can be better created and distributed using decentralized renewable energy and advanced batteries. I will revisit such potential crypto-enabled answers and contribution further below, when analyzing Blockchain solutions and applications.

No Win for Governments

One alternative seems to quickly reveal itself as a losing proposition: trying to catch up by doing what most of the industrialized world did. Most of today's rich societies directly or indirectly taxed their citizens and had the governments invest in infrastructure over decades. Looking at the numbers, basic math shows that it may take at least 50 and up to 100 years of investments to reach a comparable result for the vast African continent with its growing population.

Even assuming everything going well, this is way too slow to address the huge demands rooted in an overall very low level of economic development. Because technologies keep growing exponentially all over the world, not benefiting early on from them may even mean missing the last window of opportunity Africa has to join the world as an equal.

A timespan in the dimension of a century is not suitable for our goal of getting as many young brains as possible to work quickly, making money for Africa. Insufficient infrastructure, therefore, remains the root of much of Africa's economic weakness and a lack of opportunity for its people.

The big challenge for governments is prioritization due to a shortage of money. Without infrastructure, it is difficult to conduct any economic activity. In a globalized world, this makes it very difficult, in many areas impossible to compete. But infrastructure is expensive, and it does not put food on the table. It is a quintessential chicken and egg question.

Priorities are important. They should be derived from a look at the future. The future is not determined by raw materials or agriculture, but by technology and data-based services. Since Africa has the brains, its focus should be on establishing the best infrastructure for those to contribute, requiring finance, communication and computing, and education.

Combined, these three make up the core infrastructure components of the future, and the pathways connecting Africa's most valuable resource (its people) to their main customer base.

A Three-Step Revolution?

Letting everything sink in, our prospective African Heroes should pursue much more focused and revolutionary approaches to significantly speed up building infrastructure. In that, Blockchain plays a critical enabling role.

The solution has three parts:

1. **Focus on the Needs of the Brains.** Funding future-oriented infrastructure needs to focus on that future, and the real value it creates. This means the core investments have to only cover the needs of the 1% and maybe up to 5% of the population that initially is going to provide the quantum leap that multiplies African GDP. These investments cannot try being equitable by immediately including the other 95% to 99% of the population. The goal is to create the biggest possible impact on overall African GDP, by marketing and delivering high-value technical services for global digitized supply chains. Then, one can (and has to) broaden the base and expand infrastructure to the masses.
2. **Use Private Investment.** The directly relevant infrastructure investment has to be *complementary* to the government, not use public money or compete with it. It cannot take funds away from budgets concentrating on the

urgent needs of the other 95% to 99%. Funding via private investment enables infrastructure cost to be included in the overall business case. It adds strong and direct economic motivations to all stakeholders, without a need to artificially interject governments as intermediaries. If private investments make sense, risks are controlled, and profits can reasonably be expected, money will flow.

3. **Concentrate on Communications and Computing.** Investments need to center on such aspects of infrastructure that are critical to get African brains working as an integral part of the economy of the future. Above all this means investments into communication and high-speed data infrastructure, edge computing, 5G, and also enhanced skills training.

Blockchain-based tokens can link the costs of investments to revenues and ensure that proceeds are spread across all (token) stakeholders. Since the infrastructure would be directly built as part of a private token-governed endeavor, though, it would initially and just by itself provide a public good, although not an all-encompassing one. That would happen later.

Big City Lights: Africa Boost Centers

Half of all Africans live in urban areas and the United Nations Population Division says the percentage will rise to 75% by the year 2050.[70] These are the lowest hanging fruits that Blockchain initiatives need to focus on to have a maximum effect on African economic output.

Cities already have plenty of smart and skilled young people, better transportation systems, reasonably enough energy, clean water, food, and health care.

With sufficiently many educated brains and general infrastructure in place in a concentrated area, any solution must start with a focus on metro areas. As privately funded special economic zones, such

Technology Boost Centers, or in this case *Africa Boost Centers* (ABCs), enable focused and relatively cost-effective investments into additionally needed infrastructure aspects.

For example, low earth orbit (LEO) satellites can provide tight-beam high-bandwidth but low-latency Internet connections to such centers without having to share it with the general public. By using it only to deliver high-end technical services, a comprehensive business case can be developed. Costly 5G infrastructure, edge computing, and data storage capabilities can stay concentrated, initially to be expanded only as the growing business requires.

From there, in a second step, this high-end technical communication infrastructure can and has to spread out to other neighborhoods, where spill-over effects are likely to trigger the creation of new complementary and derivative services and products. For example, failover capacities for Internet access can eventually be made available more widely. Additional training and even energy generation can expand beyond such boost centers, and unused edge computing capabilities may be usable for other needs, like gene sequencing.

As fairly compensated African brainpower participates in building the future of the world, ever more diverse future-oriented skills spread over an increasing part of the African population. Governed smartly in a decentralized way, developing such privately funded infrastructure makes economic sense for everybody. Besides boosting African GDP numbers, society in general wins big time.

With privately funded future-oriented infrastructure, a major obstacle can obviously be overcome. But can it be done for the whole continent, or just for individual countries or a small sub-section of it?

Chapter 9

Fragmentation. Fractured, Not Decentralized

MOST OFF-THE-SHELF WORLD MAPS are Mercator projections or variations thereof. They vastly exaggerate the size of landmasses close to the planet's poles and understate the size of land close to the equator. Africa looks like it is about the size of Greenland, although in reality, it is 14 times as large.[71]

It is easy to underestimate the actual dimensions of the African continent. The East-West distance from Mogadishu in Somalia to Dakar in West Africa is almost the same as the range from Cairo in the North to Cape Town in the South, in both cases around 4,400 miles or about 7,500 km. Air travel takes in each case about 8 to 9 hours, if one can find a direct connection.

A continent of such vast size, surpassed only by Asia in population and land area, cannot effectively be approached in one big swoop. It requires engaging with dozens of sovereign nations, and hundreds of metropolitan areas, languages, and cultures. This is a monumental task by any measure. When weighing it against economic rewards, it is easy to see how profitable business cases quickly evaporate.

African Geography: What World Maps Don't Show

Fragmentation constitutes the third major challenge for an initiative to unleash millions of skilled and capable African brains on the world economy. It also does, however, point at where the greatest economic value of African empowerment can come in.

Without commercial integration, economies will remain weak and susceptible to outside influence. Investments will continue being difficult, complex, and costly, with low rates of return (ROI). The ease and scale of integration would make engagement much more logical for technology companies, manufacturing corporations, universities, non-governmental organizations (NGOs) and supranational institutions. Engagement would become more efficient, and markets would be bigger.

Today, though, one of the most dominant characteristics of Africa is diversity and fragmentation, not unity.

> "Everybody needs to go to Africa at least once in their life. If you have not seen Africa then you have not experienced the world in its fullest; then you are missing out on something that does not exist like that anywhere else."

With these words, a friend from New Zealand described Africa, many years ago. He was not just referring to the picture of himself, crouching next to a Mountain Gorilla, taken about a decade earlier. What he described was the expansive and diverse continent he had crossed, full of extremes and contradictions. The cradle of mankind. Poor but also overflowing with natural resources. Full of life, creativity, and energy. And its people!

Africa's diversity can be a strength, but much of the world does not see this. Africa is just one big blur of stereotypes. Bringing Africa to the forefront of people's minds requires both stressing the power of its diversity, but then also unifying the people on the continent to a degree that it will be recognized as an integral and equal part of the world, a part that really matters.

A Diverse Continent – Fractured but Not Decentralized

Few African countries are homogeneous. According to one study,[72] the world's 20 most diverse countries are in Africa. This has a unique impact, although on a large scale there are similarities to other continents and countries. The U.S. population consists almost solely of immigrants or descendants of immigrants from all over the world, mostly from dozens of European countries, plus descendants of African slaves, and more recently immigrants from Latin America, Asia, and also Africa and the Middle East.

Canada and Australia are similar to the United States in that respect, according to some measure even more diverse. India has hundreds of ethnic groups. Even most European nations are turning into ever more diverse societies.

Nevertheless, divisions between ethnic groups affect societies in Africa more deeply than in most countries on other continents. Elections often seem more based on traditional tribal allegiances than in the majority of nations outside the continent. Where in Western countries most of the ethnic diversity is based on immigration, this is not as much the case in African nations where different groups usually have shared the same living space for a long time.[73] The main roots for this peculiarity are decisions during the colonial times when European rulers arbitrarily drew borders as they pleased.

Ethnic diversity has been the source of conflicts, exacerbated by the fact that since the great African independence movements of the mid-20th-century, violence inside and between many African countries has been common. With more than five million killed, the globally most deadly war since World War II was fought in East and Central Africa, following the Rwandan genocide. Even then, decisions by former colonial powers and their continued meddling prolonged and amplified the severity of that conflict.

Francophone and English Africa

Still today, the remnants of colonialism are visible all over Africa. Big pieces of the West and North are dominated by the French, while the East and South show much more British influence. As one consequence, across most of the continent it essentially takes two languages to do business with Africa, with English being more widely spoken overall. Arabic, Igbo, and Amharic are probably next, and then dozens and hundreds of other languages and dialects, mostly native ones, and many related to each other.

For more tightly integrating Africa with the rest of the world it is helpful that English is being spoken widely. Communication is possible and can even be easier than with many Asian societies. Blockchain ventures can directly benefit from that, as do technology companies wanting to do business on the continent.

More important, it checks off a critical condition for doing business globally. English being spoken is good for the ambitious vision of unleashing 800 million young African brains on the digitized supply chains of the world of the near future. Although Mandarin is catching up, English still is the language of commerce.

While English, or English and French, may be sufficient to do business with Africa, succeeding in Africa will require to dive deeper. It is one reason why outsiders' approaches will encounter natural barriers when trying to define and bring solutions from abroad. Today, even Africans struggle with finding common ground.

The Pan-African Dream

Pan-African unity has been a dream at least since the colonies gained their independence from European powers, mostly around or shortly after 1960. For the longest time, this has been a mostly political goal, and it had little going for it beyond a common history of colonialism.

Africa consists of 55 nations, loosely associated under the banner of the African Union (AU) as the continent's major supranational body. The AU is nothing like the European Union, at least not yet. It has no single market, common currency, pan-African elections, and joint parliament or laws. To date, its main contribution was to be a body for coordinating continent-wide activities and initiatives. Some of these comprise a subset of member nations, or regional groupings.[74]

Among them is Smart Africa, with 24 member nations as of the middle of 2018. Its main vision is to use technology to transform African societies. However, with this being a mostly politically driven process, tangible results take time. Arguably, the biggest benefits so far have been the creation of awareness and bringing together of powerful entities and influential individuals that can help to break down barriers and to promote economic growth.

While Africa's political unification has made minor progress, economically the continent remains severely disjointed. The main reason for the lack of integration is simple: logistics, particularly relating to transportation and technology. Africa is way too expansive. Infrastructure is not just a challenge inside African countries, but even more so continent-wide. On some levels, due mostly to its size alone, building the required infrastructure is much more challenging than in Europe or even in the main population centers of Asia.

With little cross-border infrastructure, it can hardly be surprising that there is little trade among African nations. As of 2016, intra-African exports were about 18% of total exports, compared to 59% in Asia and 69% in Europe.[75] Intra-African customs are also seen as way too high and complex. To change this, in 2012 the AU agreed on the goal to establish an African Continental Free Trade Area (CFTA) through a liberalization of trade practices. In March 2018, it was eventually signed by 44 nations. The initiative is still in its early stages. With Nigeria and South Africa, the two largest African economies have yet to join as of October 2018.

Ambitious goals of integration look good on paper, but even a complete abolishment of all tariffs could not quickly solve Africa's major underlying issues. Transformative progress also requires aggressive economic growth. A still unfulfilled condition for that is the ability to freely and easily do business including sending money and conducting commerce uninhibited by governmental interference.

Even "just" improving the African Internet and computing backbone can contribute substantially to strengthen economic integration. Unless such issues of insufficient pan-African infrastructure are resolved, integration on a level comparable to the EU, or even just to North America or the Association of Southeast Asian Nations (ASEAN), seems out of reach for some time.

Is Blockchain-Based Integration "Real" Integration?

Above all, real integration has to be future-oriented and focus on enabling the free flow of data, money, and contractual arrangements, uninhibited by regulation, fast, cheaply, and reliably. High-speed Internet access, likely facilitated by satellites and 5G relays, combined with a Blockchain enabled finance and smart-contract-based commercial platform would be a major step in that direction.

Blockchain's most valuable immediate contribution to Africa can be to connect and integrate people and their business activities across the continent and with the rest of the world. Today, Africans are not only separated from each other physically and economically. As easy as it can be to pay for groceries in the streets of Kampala with a featureless phone, as difficult, costly and slow can it be to send money to neighboring Congo, or receive payment from Senegal or even the USA.

The practical implication is that such integration via a data, communication, and commercial infrastructure would counter the fragmentation of the continent without challenging the diversity of its peoples, cultures, and even nations. As a common platform, it would effectively create a single market for that economic sector which has by far the largest potential for significant economic growth and overall contribution. Integrating such capabilities into digitized global supply chains would become leapfrogging at its best, directly into the future of global manufacturing and electronic services.

We can reach such a situation just with considering a few parameters.

Diversified Decentralization: More About ABCs

Just a few small, but very significant, twists can turn fragmentation into decentralization. Three main conditions need to be considered.

- **Focus on the Blue Ocean.** Above all, the solution needs to focus on future technologies and new business models. This maximizes value creation and minimizes friction with established interests. The smaller the initial overlap to existing industries and business models, the more it turns it into a genuine "blue ocean" business strategy. Concentrating on international markets that nobody accessed before benefits everybody as a highly *additive* activity.

It is neither a zero-sum game nor does it threaten interests or even regulatory regimes that need to be overcome. Because there is value, and revenue where there was none before, this creates few economic losers.[76] In very simple terms, society is massively better off, without any (or many) losers.

- **Concentrate on Innovation and Education Hubs.** There are hundreds of African innovation hubs, over 400. This number may soon rise to be 1,000. Many of them should be turned into or complemented by the previously mentioned Africa Boost Centers. Initially, the hubs may be natural physical locations for Internet access, office space, and computing tools, plus the location for training. Since they would not have control over the commercial transactions, which would be governed via a Blockchain mechanism, they could stay independent.

  Associated with ABCs as key beneficiaries are the individuals delivering solutions, the African brains. They essentially create a professional community of high-end technology workers and entrepreneurs with shared interests and similar and complementary skills. This establishes a medium-level organization not too dissimilar to the "mediators of individual data," or MIDS, advocated for by Jaron Lanier and Glen Weyl in a recent article in Harvard Business Review.[77] For example, an ABCs can establish quality standards, use Blockchain for data provenance, and has an interest in pursuing collective bargaining. Individuals would be disproportionally empowered if all this happens in competition with other ABCs.

- **Connect Everything with a Decentralized Blockchain-Based Platform.** A Blockchain-enabled platform aligns interests plus the decentralization of governance across all stakeholders, whether investors, innovation hubs, or the African professionals providing services. Decisions can be taken quickly, including the validation and confirmation of transactions, and be guided by built-in motivations that maximize the common good of value generation and distribution. New stakeholders can then join in by buying coins. With 250 to 500 innovation hubs, or ABCs, voting will be quick while (at least in theory) economic rewards and penalties can encourage competition and also ensure cooperation for decisions that strengthen the overall solution.[78]
  The platform needs to be able to use one common payment and accounting mechanism, or currency. With it likely being a crypto-coin, though, backed by new value created as Africans conduct commerce and deliver digitized services and products, it would be subject to governance by stakeholders only. The coin then can reward investors, service providers, and individual or corporate value generators, using simple and transparent economic motivations.

The common coin may be purely technical, or it may be linked as a stable coin to another currency, a basket of crypto-coins, and fiat currencies.[79]

This can be achieved via mathematical rules determined by a community. Inside Africa, the near-ideal community would consist of the innovation hub-based ABCs, and the trained professionals and entrepreneurs associated with them.

Decentralization, No Longer Fragmentation

The above three parameters and related technocratic approach would achieve a number of significant goals. Above all, they connect and create a joint technology and economics-driven community. Since this collective is of a technical and mathematical nature, there is no real limitation for its reach. It can go beyond Africa into investment communities and to technology partners anywhere in the world.

For investors or commercial partners, and for all stakeholders, this reduces risk because it ranges across the continent's innovation, tech, and related business communities instead of traditional political centers of power. The diversity and wide spread of stakeholders make Blockchain-governance efficient, while it also reduces the ability of meddling by centralized entities.

Crypto-economics and Blockchain-governance allow for quick switches between specific service providers in hundreds of locations. This hedges against issues of weather, infrastructure, politics, and other risks. The distribution across Africa's metropolitan areas also makes training and providing services highly scalable. It makes it possible to relatively easily train, provide products for, or as customers access large numbers of locations with skilled resources.

Potentially the most important function of Blockchain-technologies is the ability to introduce mathematically expressed economic rules that can motivate people to create communities and networks spanning political and geographical borders or limitations. Africans and African Blockchain ventures can gain the most by leveraging this capability. Making the transfer of money, or of data, cheaper, easier and faster inside one of 55 countries is good. Doing the same across all of them is dramatic. Going then even beyond the African continent is transcending.

The obstacle of African fragmentation can be overcome with a strategy focused on innovation or boost centers spread across the major African metro areas. On a macro level, this sounds good. However, it is still national governments that have tax, legal, regulatory, and police authority on the ground.

Fairly or not, African governments do not have a stellar reputation. And there are 55 of them.

So we need to address that challenge next.

Chapter 10

Governments. Trusting Your People?

"DO I TRUST MY PEOPLE?" This is the question African governments need to ask themselves if they want to achieve aggressive economic development. In December 2017, I sat in the office of the Ministry of Technology of Rwanda, talking about the potential impact of Blockchain technologies on Rwanda, and on Africa in general. I phrased my main statement in the form of the above question. There are technocratic advantages for using Blockchains, but the by far biggest impact comes from their empowering philosophy.

The question asks whether African governments are more concerned about a potential erosion of their power and a threat to their political leadership, or about the unleashing of their people's entrepreneurial activity. Do they think they know best, or their people know better what is good for them? I mean these questions literally, without debating the motivation behind such positions.

"Do you trust your people?"

Fact is that governments have power, above all the ability to create and enforce laws and regulations. With that power, they can facilitate and enable. But more challenging for a pan-African endeavor to unleash new economic capabilities, they can also turn into its fourth major obstacle to be overcome. Governments can feel threatened and then may try to influence, benefit from, delay, or even stop important developments.

In comparable situations, the big economic success stories since World War II often saw governments encourage, enable, and accompany private economic activity. Aiming at similarly rapid or even faster development using Blockchain-based economic models that leverage emerging technologies, this raises an obvious challenge. We need to figure out what role African governments should have, and whether they will be comfortable with it.

In any society, if the answer to the above question about trusting one's people is *no*, for whatever reason, it leads to introducing constraints. Since the world is ever more openly competitive and integrated, this is an inherent conflict. It limits opportunities and is therefore a disadvantage. The word "development" implies change, and if it happens fast then it always upends and shifts old balances.

If the answer is yes, to act upon it requires relaxing and giving up control. In stable societies, the result is economic and ideally even political freedom. This leads to a self-reinforcing cycle that motivates people to keep seeking economic advances. The entrepreneurial spirit of the people then drives rapid development. Considering the large number of young African brains, unleashing such powers seems like a huge competitive advantage, to everybody's benefit. This is a likely outcome when governments understand that economic development is additive and not a zero-sum game.

Economic and eventually political freedom is a condition for making economic advances. It is not sufficient, though, because people also need to have some minimum economic infrastructure in place, and their basic personal needs are to be taken care of: safety, health, water, food, and shelter. When people are concerned about survival, all kinds of policies and politics can get difficult. For those in power, this then can easily inflate their sense of being indispensable. Sometimes they may even be right.

However, as dominating as their relative position may be inside the small economies of African nations, on a global scale the absolute size of African governments is very limited. Each one of the Top-4 technology companies in the world has a bigger budget for research & development (R&D) than any but nine African governments.[80] [81] In other words, 46 African nations have a smaller national R&D budget than Apple, Amazon, Alphabet, or Microsoft.

These are severe limitations. African governments need more money. Considering their increasing populations, the 3% to 5% of economic growth that Africa achieved during the past couple of decades is way too low. Governments can't like that. They need to find new and better ways to deliver more for their people. Much more. For that they need money. And therefore, in principle, interests are aligned with any approach and solution that can deliver rapid growth and bring stable money into their countries, particularly if it is privately funded. And even if it is based on empowering the young and skilled brains in a country.

As long as the governments trust their people.

Layers and Shades. Is Government in Africa Different?

"Good government is pragmatic government."[82] None less than Singapore's founder Lee Kuan Yew used these words as advice for African leaders. Outside the emotions of seemingly ever-lengthening election cycles, most people anywhere in the world would probably agree with this statement. For our daily lives, results matter.

But doing a good job is not always easy, or obvious. Sometimes it requires resisting the impulse toward "playing politics," or toward governmental activism, instead letting the people figure things out themselves. Like their counterparts in other world regions, African governments frequently do not have the answers. Nobody does when the questions are about a new vision of the future instead of one that more simply is modeled after other parts of the world. Newly emerging technologies often provide no or few references to base decisions on. This may feel uncomfortable but is more likely a good thing because it checks impulses to control complex developments that are best left alone.

For people from outside Africa, influenced by sensational media portraying Africa as an underdeveloped continent full of corruption, it can be an eye-opening experience to interact with African governments, particularly those parts not directly affecting distributive efforts. They do not only show the same kind of sophistication in many debates about regulation or practical applications of Blockchain of other technologies as their Western or Eastern counterparts. Such debates can also have a refreshing perspective of being down-to-earth. The potential impact on their people is just too far-reaching, much more so than in wealthier societies.

Even without considering politics, though, African governments have much more profound challenges than those in many other parts of the world. Above all these are a widespread lack of infrastructure and ethnic fragmentation. Because of both, tensions can mount more quickly. Then, the show (and sometimes use) of force is much more in your face than in the industrialized world.

Another specific African complication is that one does not just find national governments influencing economic activities, but also non-African governments using development aid, international organizations like the World Bank and International Monetary Fund (IMF), and multi-national organizations like the African Union (AU). Beyond these, big foreign foundations often set up by billionaire philanthropists play an outsized role. Their typical focus is on health, food, disaster relief, education, and building infrastructure in the form of schools, roads, bridges, and power generation.

In the rich world, governments provide these functions and services, or local or otherwise domestic initiatives. Again, the difference is due to a lack of money. And again, it is an argument for governments supporting alternative plans for rapid economic growth, because "local" should beat out "foreign." Not because foreigners' motivations are suspect, but simply because locals are closer to the people and better understand their needs.

Western-Style and African Democracy

When I confirmed my participation as presenter and panelist on a Blockchain meetup in Nairobi last year, I did so under one condition. On the day I was scheduled to fly in there had to be less than two presidents of Kenya. The opposition candidate had denounced the official election result, the second within a few months, and suggested to have himself sworn in as president also. Right before I got on my plane the situation was peacefully resolved. Despite some angry protests the country stayed overall calm. Uhuru Kenyatta was, and as of my writing this, still is, the only president of Kenya.

The reason I am telling this little story is not to point out an unusually chaotic and emotional nature of African politics, but rather the opposite. Emotional it is, but not unusually so. Politics in Europe, Korea, or America is emotional and divisive also, although admittedly less violent. A large minority in the U.S. has not accepted the results of the 2016 election yet, while others still perpetuate conspiracy theories about the 2000, and even about the 1960 election.

Not too dissimilar to what one can experience in Kenya or several other African countries, protests are ongoing in America. It still is the express goal of a large part of the opposition to bring down the American president with any political or legal means possible. Traditional and social media are full of angry, incendiary, and hateful statements on both sides of the political spectrum.

In Kenya in 2007, violence after a disputed election result caused the death of over 1,000 people.[83] Ten years later this no longer was the case. The country voted in late 2017 because the Kenyan Supreme Court had declared the earlier poll as invalid. The tanks did not roll in and the opposition was not imprisoned. Instead, the incumbent and previously declared winner accepted the court's ruling and the country voted again.

My point is that there is more to Africa and African politics than can be seen through a superficial view. Rich democracies should be careful not to judge prematurely or underestimate the sophistication of legal or regulatory rulings, or the importance of procedure and law in many (or most?) African nations.

Although he formally favored democracy over authoritarianism, in the same speech quoted above Lee Kuan Yew even rejected the notion of democracy being a necessary condition for early-stage economic development. Rather he said, "What is necessary for growth and progress is good government, one that is honest and effective and works for the benefit of the people."[84] Breaking this down into specific action steps, such thinking opens the door toward new kinds of governance, affecting important traditional functions of governments.

One big aspect of this is economic freedom and decentralization of power.

Governments and Economic Success: USA, China, and Singapore

Somewhat ironically both America and China are examples for the successful pursuit of liberalizing policies. In America, most areas of society were and still are decentralized. The central powers of the federal government are largely frowned upon, and its 50 states and thousands of local municipalities are to a large degree autonomous, self-governing, or at least self-regulating. People are encouraged to pursue happiness at their own terms.

Already at the time of America's founding, when the pace of technological development was much slower, the result was profound. It led to the rapid emergence of a former British colony to become the economically most powerful and subsequently militarily strongest human society that ever existed on Earth. With that, America also became a political and economic role model for most of Europe and numerous other nations, many of which followed a similar path.

Even considering its centralized and authoritarian political system, most of China's development since the 1980s also shows how encouraging the entrepreneurial spirits of individuals can have dramatically positive effects on poverty reduction and wealth creation. Like in other Asian societies the government was concerned about keeping political control. It did not, however, conclude that it had to keep a lid on economic activity. China was a poor and economically stagnating nation between 1945 and 1980. It then reached middle-income levels to a large degree via a policy of economic liberalization. We will see about the next stage, and whether the country can bridge the last major gap to become a genuinely rich society and not just a very big and powerful economy.

A Singaporean cab driver once shared with me his personal perspective on the effects when his country developed. He experienced how Singapore's founder Lee Kuan Yew led the country from a violent backwater flush with crime and ethnic conflicts by radically prosecuting all kinds of crime or illegal activity. In Singapore's case, such "crimes" also included a challenge of the underlying political system and its leadership. Beyond that, however, anything else was fair game – and so people turned to business. What kind of success story grew out of it! And once it got richer, Singapore then ever more morphed into something much closer to a Western-style democracy than it was before.

All three examples show that the position of governments can get stronger during times of rapid economic growth. But – what does any of that have to do with Blockchain?

Enter Bitcoin and Blockchain: From *Government* to *Governance*

Blockchain has been a worldwide success story despite governments and not because of them. As indicated in the Bitcoin White Paper, published in 2008, and explicitly documented in its first mined block, the creation of Bitcoin as a "Peer-to-Peer Electronic Cash System" even seems to have been directly triggered by failings in the government-dominated finance sector.[85]

Many of the Blockchain platforms, tools, and applications subsequently built were and still are motivated by a libertarian-anarchical philosophy. They often explicitly and directly challenge traditional roles of governments and big organizations. Neither governments nor banks or traditional venture capital (VC) funding facilitated the initial rise of Blockchain technology and applications.

Most definitions of Blockchain technologies define them as decentralized. This is critically important because certain functions traditionally taken care of by centralized government agencies can arguably be more efficiently delivered by decentralized, self-organizing markets. Energy, health care, data communication, and education are examples for this. Blockchains can then play an important role in trying to achieve that.

The probably most visible example, though, is the use of money. The issue is that "fiat money,"[86] the money that the general population considers "real," is artificially created via a centrally controlled process dominated by banks and politicians. It is therefore subject to manipulation by these "middlemen," or intermediaries, to their own benefit.

Decentralized cryptocurrencies can overcome such limitations and provide a greater social good for people. At least in some areas, Blockchains can straddle this more efficiently than fiat money can, particularly when the fiat money is not stable because it is not backed by strong economic activity.

This makes a good argument that Governments should give up monopolizing control over money. As long as a few conditions apply, cryptocurrencies cannot make things worse than the many weak fiat currencies. For Africa, this is particularly relevant.

Blockchain in Africa: Better Governance, Better Governments?

Blockchain-based communities can replace several roles traditionally taken on by political governments. This can result in improved services. Most notably the creation of a stable freely usable commercial and financial platform would empower people. Using new investment models for funding and developing infrastructure would also enable their economic contribution. There is more, though.

Blockchain-empowered African heroes can develop and operate solutions in parallel to how governments today fund and deliver education, energy, transportation or communication infrastructure. This should not threaten genuinely caring governments, because it is complementary and additive. There even are precedents, in some NGOs and aid initiatives providing services that in the rich world are predominantly offered by governments. Blockchain-based solutions can do the same, but with baked-in governance by stakeholders that more directly involves "the people."

Surely, governments are more sympathetic to their own people driving such initiatives, rather than foreigners, aren't they?

A key goal of governments is to establish and enforce economic and social practices that ideally benefit all members of society. In economic theory, this is sometimes called "total surplus maximization." If pursuing that goal creates not just winners but also losers,[87] then societies use governments to distribute wealth. African Blockchain solutions can take this one step further. To align interests, governments may become minority stakeholders in Blockchain ventures. In principle, this can happen in a couple of ways.

Economic growth and value generation stimulated by Blockchain motivations and shared among a large portion of a society will be a benefit for any government. Their budgets get relief as many people's wealth increases, and revenues rise even without changing laws and rules. Without specific interventions and investment, this achieves one of the goals that any government has, which is enriching its overall economy and as many citizens as possible.

Governments may become stakeholders through the equivalent of in-kind or financial investments, compensated via coin ownership. Continuous coin distribution and value appreciation of tokens can then provide tangible value to governments also. With that being the case, they can see themselves as representing societies general shareholders, and act accordingly. Like with other stakeholders, as long as they don't directly obstruct the underlying Blockchain platform, everybody is then aligned.

What happens, though, if governments don't think they can trust their people? If they wanted to stop anybody inside their nation's borders to participate? This turns into another argument for a pan-African or global solution. Due to competitive pressure, those governments are likely to lose out. Money will flow to other "nodes" in the system, to different innovation hubs or investment objects, and benefit citizens of other (African) countries.

Most people I talk to are full of sarcasm when talking about corruption and incompetence of their politicians or even governments. Adjectives like "corrupt," "self-serving," "abusive," and "criminal" frequently enter the conversation. Personally, I believe in most politicians and government employees being idealistic and altruistic on some important level. Even if I am wrong, the built-in (crypto-) economic motivations of a decentralized Blockchain-based solution that spans polities will win out in most or all cases. The economic success of the community will then entice laggards to join in after all.

Governments have all to gain and little to lose, except their underlying ability to control, constrain, and corrupt. If they want to do neither of these, Blockchains can become "good government enforcement mechanisms."

Lee Kuan Yew would like the sound of that.

Chapter 11

Funding.
From Opportunity to Investment

BEFORE I SPOKE to CK Japheth at Ntinda's The Innovation Village in Kampala, Uganda, I had arrived in heavy Kampala traffic on the back of a "safe boda" ride share. At the modern office complex, I was meeting with Olga Kiconco and her business partner Rachael Aanyu to talk about their startup Pesanest (now called Quzacredit) and about Blockchain.[88] Halfway through our meeting, CK joined the conversation and we dove deeply into challenges and opportunities surrounding technology startups in East Africa.

CK calls himself a "Startup Ecosystem Builder." He had founded The Innovation Village and was confident about the skills and capabilities represented at this innovation hub, incubator, and co-working space. In a couple of modern buildings, entrepreneurs, technical experts, and startups proved that they could succeed with the smallest of all means. Still, CK had one big wish: somebody should invest no more than two to three million US Dollars, and Uganda's probably most successful innovation center and co-working space would become an unstoppable force.

Such an amount of money would hardly be considered substantial at Sand Hill Road in Silicon Valley. It may even be regarded as too small to be worth the effort. But in Africa it is huge.

Getting Money into Africa

This is the fifth and last major challenge for an initiative to unleash the power of African brains. It is about funding, and therefore related to what our African heroes want to address most urgently in the first place: the lack of money. Blue ocean initiatives leveraging innovation hubs also need to be financed. And redesigning how people learn, stay healthy, or get energy often requires very substantial investments. Like any new venture, kicking off an initiative with the goal to generate huge amounts of money involves – money.

Currently, though, Africa has only a rudimentary investment infrastructure in place. Angel investment circles exist in the major cities, but the overall amounts given to ventures are relatively small and highly competitive. Outside some pockets like South Africa, Africa has no established large-scale private capitalization infrastructure for innovative technology projects. Things are improving, to a degree that Partech Ventures reports for 2017 a total VC investment amount of $560 million, spread over 124 startups.[89] Historically, though, there have been few deals in the $1 million to $5 million range.

So, this is now where all obstacles come together and become relevant again, with questions about confidence, infrastructure, fragmentation, roles of governments, and ROI. Investors will wonder whether in Africa, of all places, they can make their money back with profits. Above all, they will be concerned about market size, the ability to execute, expertise, and risk.

Restless? The African Diaspora.

In the early 1990s, I lived with Chinese student Cui Wu Lan in a dormitory in the German city of Mannheim. To the dismay of her German then-boyfriend, she told him that after her studies she wanted to return to China. The conversation went about as follows.

> "Why would you, why would anybody for that matter, want to return to this poor and rather backward country?"
> "You don't understand what is going on in China right now. Things are changing hugely. Everything will be different. It is a land of dramatic economic opportunities."

2018 Africa is not at the same stage. But there are indications that a similar process may already have started. Africans stay in Africa even if they have the choice for a promising future somewhere else. People with African roots "return" to Africa and foreigners settle in Africa and are spending time working with local talent to build aspirational businesses.

Like Lan a couple of decades earlier, Njoki Gichinga returned also. For the past years, she has been deeply involved in Kenya's innovation community, holding senior positions responsible for partnerships at Nairobi's renowned iHub incubator and at Safaricom Innovation Centre. Before that, though, she was living and working in Idaho in the United States when, during the process of getting one of the much-heralded U.S. Permanent Resident or "Green" Cards, she did some soul searching. Njoki realized that the dynamic environment in East Africa was so promising that she returned to her native Kenya instead.

The story of Amadou Daffe is different, but similar because he also made a conscientious choice for Africa over other parts of the world. Of Senegalese descent, he spent most of his childhood and youth in Europe and the United States. Seeing the transformation of African societies and the huge potential of educated young Africans, in 2009 he then relocated to Africa. After extensive travel, he settled in Ethiopia to start Gebeya (Amharic for "Marketplace"), a rapidly expanding pan-African IT training and programming service organization with global aspirations.[90]

Amadou is also involved with AI and IoT. He seems only limited by time – and money. Frequently jetting from Addis Ababa not just to neighboring African countries and to West Africa, but also to Europe, East Asia, and America, he incubates technology projects and businesses as well. Kukulu is one, the first 3D mobile game coming out of Ethiopia, founded by Dawit Abraham.[91] How does Amadou do it? With lots of drive and creative initiative, of course, but also with help of the African Diaspora in multiple countries.

The list could go on. Norbert Haguma returned to his native Rwanda, after having successfully worked in East Asia. Alain Kajangwe did so as well, despite his education in the U.S. opening up many doors that most people would take advantage of by staying there. Both have the belief in common that at least in the long term the combination of economic rewards and quality of life will be worth more in their home country than in wealthy nations abroad.

These are just isolated stories, but the overall numbers are big. Up to 200 million people of African descent are said to be living in the Americas alone.[92] Most of that is for historical reasons, including slavery.[93] While it is no longer a one-way street, though, a net brain drain or skill drain still seems to continue, away from Africa. Between 2010 and 2017, about one million people from Sub-Saharan Africa emigrated to Europe, and over 400,000 to the United States.[94] Over 9 million migrants from Africa are living in Europe, and over 4 million in Asia, and over 2.5 million in North America.[95] Not all is bad about that. They are an important source of know-how, knowledge, and increasingly also capital for helping the countries of their original descent.

The African Diaspora provides a knowledgeable interface between Africa and the rest of the world. It can be an important pillar for the ability to execute any ambitious business plan. Our prospective African heroes can rely on countless Africans living abroad. The interest I encounter is all-encompassing, and often emotionally charged. A conversation like the one involving Lan 20 years ago is imaginable in today's context, relating to Africa instead of China.

The result may also be the same.

There Always Will Be Muzungus

Brazilian-born American Ben Goertzel from SingularityNET started Icog Labs as the first AI research center in Ethiopia. Irish-born Ian

Grigg from Block.one and EOS launched the Blockchain-based digital and social identity solution company Chamapesa in Kenya. Finish-born Helinä Piik founded Better World Initiatives Africa in Nairobi. Formerly a VP at Morgan Stanley in New York, Sandra Ro of UWINCorp and the Global Blockchain Business Council now actively supports Blockchain projects in Cameroon and Mauritius, among others.

These are examples of entrepreneurs born and raised outside of Africa who keep spending ever more time and money on the continent or are partnering with Africans because of opportunities they see and the skilled talent they can work with.

Africans call them Muzungus, "aimless wanderers," and often are suspicious of their motivations, attitudes, or level of understanding and commitment. In a globally connected world, though, they make crucial contributions. Muzungus and the Diaspora may be most effective when working hand in hand with each other and with local experts and entrepreneurs. Somewhat idealized, they reflect the flip side of the same coin, a world fully integrated to everybody's benefit. As much as the African Diaspora links the world to Africa, the Muzungus provide the link of Africa to the world. Both open doors to know-how, money, and markets.

They are not just individuals. As an African peculiarity, large and often well-funded external non-governmental organizations (NGOs) have a significant impact in Africa.[96] Many are traditional philanthropies, frequently founded by billionaires that want to give back to society. Among the most prominent ones are the Bill & Melinda Gates Foundation, the Ford Foundation, and the Chan Zuckerberg Initiative, pursuing a variety of initiatives like improving health, providing access to clean water, and triggering social change to reduce poverty.[97]

They often take on roles that in many rich countries traditionally are taken on by governments. This establishes an interesting precedent. Like privately funded NGOs, technology and Blockchain-driven ventures can offer alternative ways for social change and economic development. This can open the door to partnerships.

Not having a well-established infrastructure and VC culture in Africa does not mean there is no interest. Increasingly, non-African VCs realize that they can benefit from investing in Africa startups.[98] African creativity is obvious. In his Pulitzer-prize winning *Guns, Germs, and Steel*,[99] Jared Diamond uses the example of Papua New Guinea to explain how a highly volatile, unpredictable, and often dangerous environment can stimulate intellect and mental flexibility. Similarly, Africans generally know how to work and function in a rapidly changing and unstable environment, under high and even existential pressure.

Traditional VCs, Blockchain foundations like the Ethereum Enterprise Foundation (EEF), EOS VC, Stellar, and Hyperledger, and large manufacturers can benefit from access to skilled and high-tech enabled African brains. Alibaba's CEO Jack Ma recently mentioned that Blockchain-based technologies, with AI and IoT, will drive the fourth industrial revolution.[100] Besides this validating the basic approach outlined herein, it also shows that the multi-trillion-dollar manufacturing industry is a potential partner in the vision.

Geopolitics and Development Aid

During his opening remarks of the 2018 Transform Africa Summit in Kigali, Rwandan President Paul Kagame mentioned that, unlike Rwanda, the little African country of Djibouti is "not forgotten for geographical reasons."[101]

He referred to the geo-strategic importance of Djibouti's unique location at the horn of Africa. The former colonial powers, plus the U.S., are not the only ones active on the continent. Beyond economics and raw materials, there are other reasons why large powers stay interested in the continent. For example, the U.S. military is active in several countries, usually battling extremist Islamic groups. Besides that, it is mostly health and food security issues that motivate America, like help to fight AIDS or Ebola.

The big new outside player is China. As part of their "Road and Belt" initiative and so-called "String of Perls" approach, the Chinese have an outspoken strategy to integrate markets across the world into a Chinese economic and political sphere of influence. Much of that is focused on making logistics easier to send African raw materials to China and Chinese goods to Africa. For this, the country offers extensive loans to fund infrastructure projects.

China clearly is seeking raw materials, but the country also actively tries to create loyalties and continuous dependencies, mainly via "help" in the form of loans. According to China's President Xi Jinping, the country is "to transform itself into a mighty force that could lead the world in political, economic, military and environmental issues."[102] There is a clear government-driven geopolitical strategic dimension in that also, executed via the one resource that China has more than almost anybody else, and that Africa lacks – money.

For better or worse, no longer are the purse strings controlled solely by the rich democracies of the world. And the motivations now also increasingly include geopolitics, rather than the post-colonial West's formal focus on helping the poorest. In response to the Asian giant's activities, the American administration has authorized a newly created aid agency to provide loans in a similar dimension as China.[103]

Although Chinese development aid reaches levels similar to the U.S.,[104] the West overall still dominates government-provided assistance by far. Beyond billions of private donations, industrial countries and supranational institutions like the World Bank annually give tens of billions of US Dollars in government-provided direct foreign aid to African countries.[105] [106] The World Bank even directly promotes the use of Blockchain-based solutions, witnessed by the August 2018 launch of a first fully Blockchain-technology based bond.[107]

The long-term impact of much of governmental development aid keeps being questioned.[108] However, these numbers show there is a lot of goodwill, and hard cash, to be tapped in. It just needs to be done in a sustainable manner that transcends traditional dependencies.

Blockchain-Enabled Investments: From Dependency to Cooperation

Looking at all the above, African businesses have plenty of potential sources of capital, even compared to the industrial world. If interest and generic availability are not the issues, what is then?

The challenges are not as much the ability to come up with creative and technically advanced solutions. It probably comes down to risks, the difficulty of private investments to get good returns, and logistical complications. These are important for investors everywhere. Acceptable ROI and risks require markets of sufficient size and diversity, and for African products and talents to get access to global customers.

Blockchain-based approaches make crucial contributions to each of these three areas.

- **Scale.** They facilitate scaling out. One of my main arguments is that a common Blockchain-based platform can establish a self-coordinating and cooperative network of innovation hubs, or technology "boost centers." By transcending borders and connecting innovation hubs across the continent, they can deliver the numbers for both demand and supply.

  As a result, markets for investments into critical infrastructure increase in size. They become pan-African, and solutions can be pushed across the continent using one network and platform. This also allows playing a "numbers game" by hedging risks across a number of societies. In the other direction, it enables investors, customers, and partners to access innovative and entrepreneurial talent no matter where it is on the continent. No longer do they need to go nation by nation.
- **Aligned Interests.** Blockchain-solutions can align interests over a large and diverse set of stakeholders. Measuring newly created value in tokens shifts traditional roles, creates more of a community, and promotes a cooperative approach. Tokens can link investments into education to the providing of services

using these acquired skills, or investments into a high-bandwidth Internet connection to its use for professional services. They also can link capital investments into gene sequencing to treatments of sicknesses, or building renewable sources of energy to a distributed pan-African market of electricity. An important side-benefit is that this prevents lopsided relationships between foreign lenders and local entrepreneurs, or middlemen corrupting the use of investment funds. Entrepreneurial activity becomes a more genuine joint effort.

- **Global Appeal: The Internet of Brains.** Blockchains open the door to foreign markets, by integrating African skilled resources into an Internet of Brains (IoB). I define IoB similar to the IoT. It is a decentralized and self-organizing global network of people acting as a platform to deliver skills-based digitized services to customers and consumers. This further increases the market size for services and solutions, even dramatically so when compared to African national economies. With foreign customers paying for services, it reduces risks of investments that develop and empower African technically skilled resources and entrepreneurs. Returns will not be earned in small African economies but in the industrialized world.

Investment and Trade, Not Aid

Using a token can align interests of diverse stakeholders, mitigate risks, bridge borders, and provide scalability. Crypto-economic motivations can align interests and contain a decentralized but still effective governance mechanism, including the ability to fine-tune the initial incentives.

Tokens then fund investments, which bootstraps and launches profitable ventures – and generates even more money for Africa. They do so in the ownership of a diverse set of stakeholders that make high-value and profitable contributions to Africa and the world economy. This money will not be controlled by a federal bank, but by stakeholders who have a common interest in maximizing the value of the coins. Everybody can benefit. Value that otherwise may not be created at all, or would remain with the industrial world, owners, or lenders, will at least partly remain in Africa.

So much about my personal theories. It would really be interesting and helpful to know what Africans think. And so, I decided to ask them.

Chapter 12

The Real Experts: What Do Africans Say?

KLAB IS RWANDA'S biggest technology innovation hub. It is run by Aphrodice Mutangana, whose activities I somewhat poetically liken to "fertilizing young brains." Like many Rwandans, this modest but driven former programmer draws a lot of motivation out of his personal experience of the country's genocide a couple of decades ago. With his own personal crowdfunding initiative, he actively supports women that survived the genocide. In his day job, though, Aphrodice has an even bigger impact on his country's future. As kLab's General Manager he facilitates the education and training of thousands of Rwanda's young, fostering their technical skills and entrepreneurial spirits.

He knows how to reach and jump-start these young brains and minds, and so he and I decided to co-organize a Blockchain workshop as a side event of the 20+ nation Transform Africa Summit 2018 in Kigali. This is where I undertook a unique step, supported by two grad students of the University of Utah, Cameron Owen and my daughter Fiona Kastel. To our knowledge, up to that point, there had been no formal attempt to specifically capture the mood and thinking of informed African Blockchain enthusiasts during such an event. As of May 2018, there had hardly been any such events in Africa at all.

At *TheWorkshop*, though, we tried to also let Africans do the thinking and explaining, bottom-up. We posed 25 questions to an audience of over 200 Blockchain enthusiasts who had just been exposed to and educated in a wide range of Blockchain-related topics, increasing their knowledge level and understanding.

Talking about Blockchain in Africa can only be complete with such meaningful input from African audiences. Anecdotal "evidence," books, Internet research, or explanations by "experts" can reveal a lot about a subject. They cannot substitute for personal interaction and input from the subject of study itself. Expressed differently, feedback from African Blockchain enthusiasts is necessary, and at least as valuable as explanations from experts who often enough are outsiders. Africans involved with Blockchain are experts in just that, Blockchain in Africa.

The Survey

So, on May 9, 2018, we co-organized the Blockchain-focused *TheWorkshop* with kLab, in coordination with Smart Africa whose *Blockchain Hub* we also supported on May 10 and 11, 2018. *TheWorkshop* included presentations and panel discussions, from Bitcoin (in the form of Bitcoin Cash, presented by nChain), African Fintech (Stellar, Eversend), solutions for Central Banks using Hyperledger (Soramitsu, Rwanda Central Bank), health care (Healid), identity (Chamapesa), and more perspectives on African solutions and Blockchain technologies (e.g. by Sandra Ro about projects in Cameroon and Mauritius, and Gebeya in Ethiopia). Norbert Haguma managed and coordinated all these activities under the banner of Smart Africa.

After the event we conducted a survey to get input from our almost solely African audience. 97% of the participants were from five Africa countries, with 88% from Rwanda. About 80% were male and almost as many were between 20 and 30 years old. Less than 30% said they were knowledgeable about Blockchain, but two-thirds said they came to the workshop knowing little or nothing.

After a full day of immersing everybody in a wide range of topics relating to Blockchain, we solicited their input, assessment, and opinion on a number of Blockchain-related topics. To motivate participation, we created a crypto-fountain, with us sending each survey participant who gave us their wallet address the equivalent of $5 in Bitcoin (Bitcoin Cash - BCH). The result was not just that we added dozens of new coin-owners to the African crypto community. More important was what we learned from our audience.

The Opportunity

About two-thirds saw the highest priority for Blockchain to achieve economic freedom (64%), reduce corruption (64%), and above all to have a means for personal savings (69%). When asking about the special appeal of Blockchains as a way to achieve these goals, only one answer was allowed per person. Half of all people liked Blockchain because of the closely related abilities to eliminate intermediaries (28%) and establish peer-to-peer capabilities (22%). 17% appreciated its open source aspect, and still, 15% were particularly attracted by its potential to help breaking down geographical and political barriers. The most notable piece of information we gained is that less than 10% found crypto-coins themselves as particularly appealing.

This is in stark contrast to what we have seen during the ICO craze around the world, even today. It is not the chance to grow assets by investing and HODLing.[109] Rather, Africans seem to recognize Blockchain's ability as a new technological infrastructure to empower people. Not surprisingly, 85% are convinced that Blockchain can help improve lives and drive social change in Africa.

To leverage the capabilities of such infrastructure, the survey respondents stressed the need for more education. Over 75% saw as the number one impediment to achieving social disruption a lack of Blockchain education. 25% to 33% each also saw further challenges due to the inflexibility of people, the lack of infrastructure, and governments. This is encouraging. It indicates that people basically said, "let us know how Blockchain works, and we will use it to improve African lives." Our participants already seem "empowered" although not yet "enabled."

We can question whether these sentiments are typical for all of Africa. That survey respondents were predominantly from Rwanda may distort the picture. Their home country has a widely acknowledged technophile government that formally encourages bottom-up development. At least, this seems to be the general perception of the Rwandan people, despite them somewhat ironically letting their government lead and coordinate much of these efforts.

Special African Conditions

When we asked what special conditions need to be considered for Blockchain in Africa, two answers dominated: mobile phones and Fintech. Neither was surprising, particularly considering the East African context. More than 40% of all participants mentioned the high use of smartphones, and 30% pointed at Africa's pre-existing experience with mobile payments.

To anybody with hands-on exposure to Africa (and particularly East Africa), these answers made sense. They confirm what one could have discovered by doing online research. Much more significant, arguably, are the low percentages for Internet speed and low computing power (9% and 5% respectively). Again, this seems to confirm a confidence in one's ability to overcome technical issues. Even more extremely, one can interpret this as a vote for opportunity over issues.

When looking at patterns in the answers, this self-confidence and positive thinking could be because of the predominant age group being in the 20s, or the workshop participants and respondents being technically trained or at least interested in tech. That would then allow us to draw other conclusions, confirming the above-mentioned competitive advantage of Africans: hundreds of millions of young brains and minds that are trained in technology.

Almost 60% of the participants considered existing Blockchains as sufficient for the transformation of Africa's social, economic, and political systems. They seem to sense that applications are more important at this point in time than trying to build yet another "African Blockchain." There is a slight contradiction, though. 28% see public Blockchains as having the biggest impact on Africa, 14% said private Blockchains would be it, while 70% see both. Since most private Blockchains would be new ones, an overall 84% are in effect stating that new Blockchain solutions would play a major role affecting Africa.

Maybe this reflects a lack of understanding about the difference between Blockchain, platforms, and applications. My interpretation is that most people seemed fine using existing Blockchain platforms but wanted new Blockchain solutions built on them. This could be positive news for the likes of Ethereum, EOS, Stellar, Cardano, and Hyperledger, which all offer smart contract capabilities.[110]

Some answers would warrant diving much deeper to get clarifications or to verify the real understanding of the respondents. The additional value gained of such an exercise may be limited, though. Workshop participants clearly stated that education is still insufficient. This may explain some of the contradictory answers.

Blockchain Features, Applications, Use Cases for Africa

The survey showed no real surprises about what features are most important for Africa. It comes down to cost and usability. A majority of 57% wants low costs, but almost 50% each voted for "ease of use" and the related ability to use with mobile devices." Performance was next with about 42%, still ahead of "smart contract capability" and anonymity.

One core element of many Blockchain-based applications is how they deal with identity. Although the communal character of African societies was mentioned multiple times at *TheWorkshop*, less than a third of the respondents see a community-based identity solution as being best suited for Africa. Most liked the use of more technological and individualized solutions, particularly biometrics (57%) and mobile phones (40%).

Not surprisingly, the verdict about Blockchain's opportunity for positive social and economic change was almost unanimous: 97% of the respondents saw more opportunities, and only 1% more risks. This was despite the fact that 86% were concerned about the safety of information on the Blockchain. Only 6% were not concerned about it. This is significant. While the attitude toward Blockchain as money was overwhelmingly positive, there is an area of pain that requires additional education or points at a significant opportunity for Blockchain-solution providers.

In rich countries, the recent ICO craze has largely been based on entrepreneurs, developers and Blockchain enthusiasts coming up with a variety of applications that go far beyond Bitcoin's goal of introducing a cashless payment system. If the survey participants really represent Africa well, there seems to be a certain "back to the roots" position.

Looking at a wide range of Blockchain application areas, Fintech with 80% and commercial applications involving supply chains and contracts (72%) dominate African priorities for the biggest positive transformative potential in Africa. Other options garnering more than half the votes were governmental services (62%), agriculture (58%), and health care (52%). Energy with 42%, manufacturing with 37%, and politics with 30% were below that but still with significant numbers.

The by far biggest use case for Africans is Blockchain-enabled intermediary-free Fintech. No less than 84% consider it either very helpful to significantly drive economic development, or even necessary for that. About one-quarter of all respondents had the opinion that real opportunity for economic advancement cannot exist unless financial markets are independent from political and economic elites. Only 3% deemed it not important.

Considering the relatively large number of Blockchain solutions already in existence, it is surprising that less than a quarter of all participants did express confidence that these existing solutions were sufficient for Africa. Almost a third saw it as necessary to create an independent African crypto-coin, more than expressed confidence in already existing Blockchain solutions. A majority of 45% stated that it depends on how existing Fintech solutions succeed. That reflects a healthy portion of skepticism.

Diving deeper into what Fintech applications are most important for Africa, a balanced picture emerges. Consumer-based financial transactions (64%), followed by cross-border transfers (63%), savings (55%), and then enablement of IoT (48%) and investments like micro-financing (40%) cover the full bandwidth of the typical roles for money. Interesting enough, though, store of value was not the dominant theme, but Blockchain's use for financial transactions.

Strategy Choices

In the final section of the survey, we tried to explore the best strategies to achieve success for Blockchain initiatives in Africa. The answer about the biggest obstacle could hardly have been clearer. It was in line with the above-mentioned biggest challenge to reach the potential for social disruption, which was close to 80%.

Consequently, "lack of training" dominated the list with 55%, far ahead of "politics" (15%) and "lack of capital" (14%). Interestingly enough, "challenges by established interests" were only mentioned by 9% of the respondents, and "outsiders" or "others" less than 7% combined.

On some levels, unusual results came out of the question about how Africans should approach the transformation of their existing economic, social, business and political systems. The number one response, by over two-thirds of all participants (69%) was to ask for connecting African innovation communities. Not far behind with 64% was to carefully and incrementally improve existing infrastructure and capabilities through education and technical training. These two points may be connected, with the innovation communities seen as the main driver for education and solution development.

Notable was "governmental support" with 53%. This differs from the anti-government philosophies that in general seem to dominate the Blockchain world elsewhere, with government seen as one of the most important intermediaries to be disrupted.

"Disruption" per se was not an option given, but "transcend and leapfrog" came close to it, garnering no more than 19% support by respondents. This was almost half of the also low "opening up to financing" and "ensuring investment."

The survey conveyed enlightening and important perceptions of the developing African Blockchain Community. They are key pieces of the "Blockchain in Africa" puzzle. Obviously, our actual and prospective heroes of African Blockchain and technology ventures know what they want and need.

Now that they told us, it is time to put the various pieces together.

Chapter 13

21st Century Solutions for 21st Century Challenges

MANY THINGS ABOUT New Zealand are beautiful. Yoon Soohyoung and Dongyeol Shin, however, were particularly fascinated by the trees. When we picked them up from Auckland airport on a crisp day in the Kiwi spring of 2015, it was not the coastline, volcanoes, and grassy hills that attracted their attention. Besides the sheep, one of the first things catching their eyes were the many old trees that most people in the world just take for granted. The reason was simple: There are very few in Korea. The Japanese occupation of the 1930s and 1940s, and then the Korean War in the 1950s had resulted in massive and near-complete deforestation.[111] Korea had been completely destroyed and its impoverished population needed firewood to cook and survive harsh winters.

By nearly any measure, as late as in 1965, Singapore and Korea were poor countries. Right after the heydays of African decolonization, Singapore's GDP per capita was $516 and South Korea's a paltry $108 (in current US$). In comparison, Ghana had a GDP per capita of $266, the Democratic Republic of Congo (DRC) $220, Kenya $97, and Nigeria $93. Only 52 years later, the respective numbers were $57,714 (Singapore), $29,743 (South Korea), $1,968 (Nigeria), $1,642 (Ghana), $1,507 (Kenya), and $458 (DRC).[112]

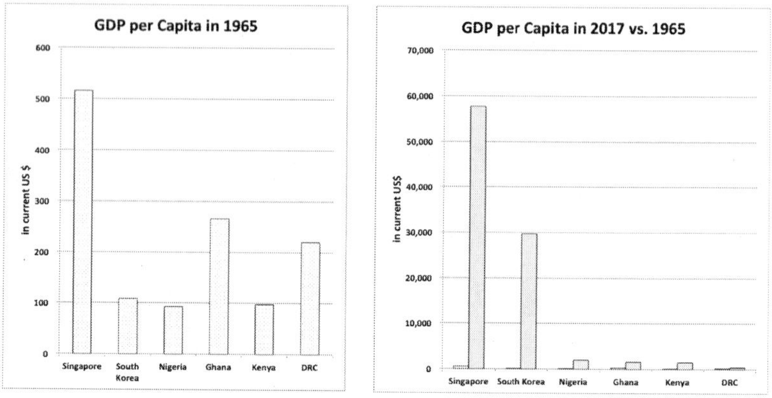

Figure 6.
Rapid Growth in Asia vs. Africa

The country of Singapore managed within 50 years to turn from one of the globally poorest countries into one of almost unsurpassed wealth. South Korea achieved even more significant growth in about the same time span.

Africa can do this also. But the lessons from recent success stories should not be to simply emulate what Singapore and South Korea did. Times have changed, technologies have moved on, and the global economic and geopolitical context is different. Today requires different tools and approaches.

What can be learned from Singapore and South Korea is that very profound change is possible when countries open up for business, and when societies consider new technologies.

This has a special relevance and urgency for Africa. Technological advances and their adoption rates increase in speed, driving accelerated change in societies. The Internet, mobile phones, and social media are examples for that.

This can level the playing field more than ever before. But it can also widen the gap between those that take advantage of emerging technologies and those that do not.

Much of Africa tries to catch up to today's industrialized world. But its economic growth of 3% to 5% since the turn of the century is too slow. Over the next decade, once the effects of automation and productivity increases via AI and the near-complete digitization of world economies fully kick in, a still poor Africa will struggle again to even just keep up the pace. If it continues on the current path, Africa may stop catching up and again fall further behind.

It becomes a question of what to focus on. If African countries continue catching up to the 20th-century solutions of the Western world, this is where Africa will end up in a couple of decades: in the 20th century. But by then we will be in the middle of the 21st century! Africans must use 21st-century tools to address 21st-century challenges. This way, they can preempt developments, stay ahead of them, or even lead. As the world evolves technologically, at least part of the continent will already be there.

Alternative Realities. From the 19th Century to the 21st Century

The roots of many of the rich world's infrastructure date back about a century, addressing issues of the industrial revolution.

Starting with the 1880s, the Western world developed its current public health insurance model, then expanding it into an ever-larger system of government-provided centralized insurances and social services. Centralized electricity grids originated in the 1920s and 1930s. The Fed, the U.S. central bank, is just a bit over 100 years old.[113] The American federal public road system also dates back about 100 years,[114] to the years immediately following World War I. It is roughly the same time that a federal income tax was established for the U.S.[115]

These solutions became hallmarks of modern societies.

However, they reflect the hierarchical thinking, technologies, and capabilities of their time. Since then, they have not been replaced or revolutionized but mostly changed via bolt-ons. Meanwhile, humanity has passed through what often is called the information age and is entering the digital age. And what seems to become ever clearer is that the old centralized solutions are full of problems, vulnerabilities, and inefficiencies. We require a more profound shift in social and economic models.

Today, the United States spends substantially more money than any other nation, in absolute and per capita numbers, on social services, education, and health care.[116] Various other countries are not far behind in the numbers, but they seem to have one thing in common: many citizens of Western societies consider them overly bureaucratic, inefficient, and delivering under par. For example, over 30% of American *doctors* say that they are spending more than 20 hours per week on administration – and patient feel the same.[117]

Dissatisfaction mounts while people are increasingly concerned about the ability to continue funding these programs. Particularly in the U.S., costs keep spiraling out of control and are the main reason why the federal debt has reached record levels.[118]

The centralized solutions seem to have reached natural limits that simply cannot be perpetuated. They need to be replaced and transformed.

But many African countries still try to catch up to them.

Making Today's World Better

Cardano is one of the biggest Blockchain initiatives, with a mid-2018 market capitalization of a couple of billion dollars. Using a strictly scientific and peer-reviewed approach, they aim at becoming the leading smart contract-based platform for the newest generation of Blockchains, essentially replacing Ethereum (and competing, among others, with EOS). Through a large part of 2018, its founder and CEO of Input Output Hong Kong (IOHK), Charles Hoskinson, pursued several projects in cooperation with Ethiopia's Ministry of Science and Technology.[119] They aim at better tracking supply chains of coffee and improving on the current system of land ownership records.

Projects like these give a glimpse of what Blockchain technologies can enable. They can eliminate fraud, lower costs, increase speed and quality control, and maybe even cut out waste created by intermediaries. Such use of crypto-technologies, though, due to being limited to one country, one industry, and traditional processes, will mostly result in overall relatively minor improvements. They keep the nature of existing solutions intact. While not insignificant, the economic value added is therefore only incremental also. Small steps can add up and go far, but they are only enough to catch up if one has enough time. Meanwhile, the divide will widen.

But Blockchain technologies can enable more. They can change the nature and rules of whole industries and drive more innovative and effective investments into key sectors of society. The sheer act of investments can then drive economic growth. Even more important is, though, what one can accomplish with full integration into the global communication and data flow, decentralized renewable energy, better and more adaptive education, and personalized predictive health care. It is safe to assume that additional GDP growth rates by several percentage points are possible.

This is still not enough. More can be achieved.

Leapfrogging Directly into the 21st Century – Just "Doing The Right Thing?"

Leapfrogging means jumping over an unnecessary step. For Africa, it means stepping directly to 21st-century solutions without first trying to catch up to by now outdated 20th-century answers. The rich nations cannot do that. They are already in the 20th century, trying to reach the 21st by making painful adjustments whose value often is difficult to communicate. Their current solutions (sort of) work, and therefore their arguments for transformation are less obvious.

For Africa, it is not just possible, but necessary.

It requires letting loose of old concepts and thinking, and daring to venture out into the blue oceans, the unchartered territory. These anticipate and take advantage of global technological and economic trends and get in early on their upsides, participating on as level of a playing field as there is. In these new waters of technology and commerce of the near future one can develop novel solutions, open new markets, and encounter and take advantage of original and much larger opportunities.

Innovation is the key word here, not copying old and by now outdated solutions. Instead of catching up to the rich world's 20th century standards, this is about going back to first principles and using Blockchain to change the way substantial portions of society work.[120] It requires developing and using new (crypto-) economic and governance lines of attack to bring about systemic change.

For example, today there are no substantial digitized supply chains in place that leverage XR and IoT. Logically, they can therefore not be improved, transformed, or taken advantage of. Gearing up for that future establishes new economic benefits. No longer is it about competing, but about anticipating and acting.

Anybody in the world can participate in these developments. The impact on Africa, however, can be disproportionally large, because of two reasons. First, in the short term, its relative poverty means that even integrating small portions of skilled Africans into this new world will have a huge impact on its economies. They will, therefore, grow much faster. In the long term, then, the sheer quantity of brains seamlessly integrated into the global economy of the future will enable Africa to sustain or even expand its relative and absolute position.

There are 12 Zeros in a Trillion

Fairly simple calculations show that the effect can quickly reach and go beyond an additional 5% to 10% of GDP, possibly substantially more. Just assume that ten years from now 1% of the African brains would actively contribute to such high-value activities and would get fairly compensated. Do the math, in U.S. Dollars.[121] How many zeros does a trillion have?

This is new economic activity, previously non-existent, but of high value and leveraging little more than African brains and modern communication and computing technologies. Investing in this future has a big ROI. And by introducing the underlying economic motivations and establishing the monetary and commercial platform,

Blockchain will be a key enabler of such investments, with the potential to profoundly accelerate and drive African development to a so far unheard-of scale.

800 million young African brains are available to join the global economy when supply chains continue to get digitized via VR and XR, Fintech, biotech, automation, personalization, and additive manufacturing. To deal with such a world requires little more than – brains. It does not matter where these brains are located, just how they contribute. This is Africa's big opportunity.

Why Digitization Is Good for Africa

The digitization and decentralization of supply chains do not bode well for Africa becoming the next manufacturing hub in the traditional, physical labor-based sense of the term. Technology trends, and therefore the numbers simply don't add up.

But inherent in these same trends lies the source for triumph. If supply chains are digitized during design, extended reality-based customer interaction, the customization and personalization of products, and via IoT also for product maintenance, then the main condition for manufacturing success becomes brains and not factories. This is similar to biotech, where genetics is enabling a food and health care revolution by focusing on the data encoded in genomes. It also is comparable to the world of finance. Crypto-enabled Bitcoin already introduced a peer-to-peer payment platform that enables the leapfrogging of traditional banking, whether for paying, saving, or investing. It even is happening to energy where the combination of new batteries and renewable energy, aided by AI, drive the creation of micro-grids.

The common denominator of these trends can be expressed as a combination of decentralization and digitization. It can also be described as organically growing trustless crypto-networks with smart-contract automation, P2P financial transfers, and instantaneous connectivity of people and things.

Or it can simply be called: Blockchain technologies.

Blockchains can enable efficient domestic and cross-border commercial transactions, inside Africa but even more importantly between Africa and the rest of the world. It can provide the platform for hooking up African brains to the future digitized economy. Even if this starts small, which it will, the absolute economic impact for Africa can be dramatic.

Africans cannot be successful by being timid, applying today's solutions to tomorrow's challenges, or by using yesterday's tools to take advantage of the opportunities of the near future. All that does is tinkering at the edges and improve some processes, but it is not transformative, and definitely not transcending or leapfrogging. Lowering the cost of a financial transaction from 1.5% to 1% by using Blockchain efficiencies is good. Storing governmental files cheaper on a decentralized database using IPFS is progress.[122] Tracking ivory via Blockchain entries is helpful.

By itself, though, all this achieves little.

Connecting African brains into the digitized supply chains of the future – now that is where the real and transformative opportunity lies. And remember, this is not just transformative for Africa, but it goes both ways since the rest of the world benefits from it also.

Diving deeper into the mechanics of Blockchain can indicate how this can work.

The Blockchain Panacea?

A panacea is a cure for all ills, resolving all difficulties. The hype around Blockchain can let one believe that it is the panacea for all problems of society. Blockchain critics use such claims, combined with common Blockchain-related challenges like cost, performance, or ineffective governance, to attack the craziness of this "fad" and "bubble."

Let me be clear. Blockchain-technologies alone will not cure cancer or end corruption, exploitation, and wars, bring universal health care or health, unlimited high-quality food supply, cheap and clean energy, while maintaining the dignity of every human being.

But, and this is a huge *BUT*: Blockchains can make major, possibly even indispensable, contributions to achieve these and other lofty and mighty goals. They do so together with other emerging technologies like artificial intelligence, extended reality, genomics, additive manufacturing, high-throughput satellites, and the Internet of Things. Inherent in Blockchain technologies are core concepts and key enablers for genuinely transformative or transcending solutions to societies' challenges. These concepts may even be requirements to leverage capabilities of other technologies. Just to name one, it is hard to imagine effective IoT without some inherent Blockchain-like consensus and accounting mechanism.

The following short paragraphs about technical concepts of Blockchain can provide context.

> On the most basic level, Blockchains are just another form of mathematics, implemented in computing networks. They are databases using math and widespread networks of computers to determine which entries are correct – instead of relying on a centralized administrator. The methods with which agreement on correctness is reached are called "consensus mechanisms." On the one side this makes them in principle very resilient against being manipulated or broken, on the other side it's requiring consensus among a large number of computers also slows the databases down and makes them costly to maintain.
>
> Although they are no panacea, it means Blockchains have a place in the domain of computing, and even for many or even most kinds of social and economic interactions. Whenever multiple parties want to document the truth of a transaction, using Blockchains can make sense (or of comparable technologies, like the IOTA "tangle").[123] For example, this truth can be to verify that a document is the one it claims to be, for any document, even if it is not stored on the Blockchain itself.

> Every time trust is required, Blockchains excel since they do not depend on third parties. This leads to their best and most obvious use cases, to document any kind of transaction between multiple parties. Contracts and transferring money are typical examples.

Bitcoin as the world's first large-scale Blockchain application was designed as a "peer-to-peer electronic cash system."[124] Various other forms of what one can consider "electronic cash" already exist, either as cryptographic solutions, mobile airtime, or credit cards. The most important contributions of Bitcoin are therefore not simply its electronic or cash components, but that its underlying Blockchain model is peer-to-peer and decentralized, its token supply is limited, and it contains economic incentives to reach a consensus about the truth.

Math + Economics = Real Power

A Blockchain's mathematically enforced economic rules and motivations can align the interests of humans (or machines) to organize themselves and their actions in novel ways. Instruments like these may be among the most powerful that human societies have yet seen. They give us new ways to establish and govern complex organizations, processes, and markets that leverage emerging technologies.

Far from the hierarchical and linear models that originated in the industrial revolution, Blockchains use targeted rules and incentives that are mathematically enforced and do not require trusting a central authority. As a consequence, networks can organize themselves much more efficiently and scalable.

The inherent power of such crypto-economic models lies in their not being limited by the imagination and direction of just one person or group. Rather they intrinsically leverage the capabilities of all network participants, coordinating their actions via rules expressed as targeted motivations. This re-introduces the natural state that we find all around us in every aspect of nature: organically growing or changing systems, self-organizing, resilient, and adaptive.

For example, in education one can imagine targeted motivations that give more tokens to somebody finishing degrees early, or acquiring sought-after skills, or using new skills on the job versus acquiring them "just for fun." Regarding energy, driving on a toll road with an electric vehicle may cost less than using one with a diesel engine. When genomic data is collected and analyzed, early contributors may receive more tokens as compensation since they incur higher costs and risks.

As new tokens get created due to economic activity, early investors, or certain network members like local communities, may get a fraction of them also to compensate for their risk-taking and special contribution. Or when a Blockchain solution collects revenue, ownership of tokens can be the basis to distribute this revenue.

Of course, all this will never be perfect. Like all human activities, engineering and designing complex social and economic mechanisms will always have shortcomings. Competition and ways to incorporate consistent improvements can help with that. Widespread formalized rules of decentralized "governance" originated in the 1780s, when the citizens of the United States agreed on the first-ever formal constitution.[125] By adopting a clear but flexible format, they established a largely decentralized nation consisting of states, a rapidly growing organization (or "organism") that lasted for a quarter millennium and weathered dramatic political, military, societal, and technological challenges.

Today's more complex world requires even more and faster adaptation. Therefore, without applying crypto-economic or Blockchain decentralized concepts of governance it may be impossible to achieve any of the above-mentioned ambitious goals of humans.[126] Hierarchies have significant limitations, organisms much less so.[127] It may not be the best example but compare your ability in a democratic nation to keep the "oh so powerful" police out of your house, versus a tribe of tiny ants. In the U.S., at least ideally, the way to achieve the first is by saying "no," while confronting the tiny insects require often elaborate, costly, and long, drawn-out efforts.

But the goal of decentralization is not to "make" things or people equal. This cannot be done since no two humans are equal. Rather, decentralization is about making every actor play *by the same set of rules*. This enables the organic behavior of people in society. Motivations are in the long run stronger than force.

Taking away power from simplistic political hierarchies or from monopolistic businesses and entities introduces balance. Blockchain-based models expand on traditional concepts like money, shareholding, or political power, and introduce new ways to bridge geography, time, decision-making, and use of assets.

The long-term success of these processes is not guaranteed, but they are promising. They reflect a more natural state of complex systems and align interests across geography, culture, ethnicity, and polity.

Chapter 14

No Longer "Survive" Instead of "Thrive"

SOME OF AFRICA'S and of Blockchain's challenges look intriguingly similar, and so do their likely solutions.

This brings me back to April 4, 2018, and the *Deconomy* Blockchain conference in Seoul, South Korea. One day before the Ethereum panel discussion mentioned in my opening chapter, the topic of discussions was how to achieve high performance, low cost, and secure Blockchain transactions on a massive scale. Staying with the theme of the conference, the solution was to be based on Bitcoin remaining "decentralized."

"Nodes do not matter, Edges do!"

More than any other statement, this sentence became my favorite and most eye-opening moment in Seoul. Craig Wright delivered it in his usual bombastic form during a presentation that triggered an emotional debate between Blockchain thought leaders – swear words, ridicule, yelling, and personal insults all included.[128]

The talk about nodes and edges also had a direct relevance for Africa. Craig Wright even had explicitly defined his goal to create a currency that was usable not just for the rich world, but for all Africans also. This struck a chord with me. Even more so, because his presentation had focused on network theory within principle universal applicability.

> In technical terms, his statement was to say that a relatively small "near-complete" network of about 1,000 competing nodes, all (or almost all of which) are connected to each other,[129] is faster, more decentralized, and therefore more secure than a much larger network with more nodes but less direct links.
>
> Is science really showing that for performance in a complex network (the number of) "nodes do not matter, [but] edges do?"
>
> On some level, of course, both are important, and so are algorithms enforcing corresponding economic incentives and penalties. Graph theory defines that mesh networks are less efficient, and more centralized than near-complete or complete networks, mainly because the number of edges to nodes (or vertices) is smaller. Near-complete networks may have smaller numbers of nodes, but most of them directly connect with every other one. This makes them faster, more resilient, and thus usually more secure.

Many recent Blockchain-related engineering efforts consider this. Some Blockchain platforms have been trying to address limitations of performance, cost, and overall scalability by reducing the numbers of actively required nodes.[130] Many shifted their focus to improved consensus algorithms, often requiring fewer nodes. EOS is such an example with its 21 block producers, and permissioned Blockchains like Hyperledger are so almost by definition. Stellar's 39 validators are just a small group as well,[131] and even the 1,919 nodes of Bitcoin BCH versus the 9,917 Bitcoin BTC nodes and Ethereum's 17,653.[132] To overcome limitations that may directly or indirectly be based on too many nodes and too few edges being involved in consensus mechanisms, the Ethereum community works on sharding, side-chains, and a shift from Proof-of-Work (PoW) to Proof-of-Stake (PoS).[133]

Some of this is just logical and inevitable, at least at some level. Tens of thousands of loosely connected mining nodes spending effort in the form of electricity to validate a transaction can hardly be faster or more secure than a much smaller number of them that are all connected to each other. The latter can reach almost instantaneous consensus.

This is not yet a sufficient condition for performance, but in real life, a few directly connected industrial-scale validation nodes just have to be faster and more economical on almost every level. The debates then focus on the possible trade-off to achieve high-performance, i.e. security and decentralization, and on how consensus is reached in a trustless network.[134]

Just like Africa?

African Nodes and African Edges

It is almost ironic how much the state of the African technology ecosystem and economy resembles some of these challenges of the major Blockchains in 2018. This is witnessed by terms like "uneconomical" or "costly," "slow," "complicated," and the debate about "central control," "trust," "malicious actors," and "risk of abuse." Depending on the criteria used, Africa has many nodes (mobile phones, computers, bank accounts, or simply: people) unequally connecting the most advanced of about a billion brains, resulting in a number of issues. It is decentralized, alright, but efficient it is not!

This lets me borrow some of Blockchain's approaches to address African economies' shortcomings. After all, the whole continent could benefit if Africa's economic nodes would be better connected, integrated, with capabilities that make direct communication and economic interaction much more efficient.

With 55, Africa has the largest number of nations of any continent.[135] Just consider this number already hints at the special challenges of Africans to connect with each other and the rest of the world. But there is more to that point. According to a recent and probably already outdated count, 442 innovation hubs exist on the African continent.[136] There are about one thousand universities or other institutes of higher education,[137] thousands of cities, with at least 65 having a population exceeding one million and 183 having more than 300,000 people,[138] plus hundreds of thousands of small and many large non-governmental organizations of which hundreds if not thousands operate internationally.[139]

All those can be considered nodes. Above I have shown how inefficient it is today to transfer money and do business in and with Africa. It leads to gross inefficiencies, high costs in cash-strapped societies, significant barriers to entry, and it prohibits technological and commercial activities. This happens because these nodes of countries, cities, or innovation hubs are not really efficiently connected. They behave like a mesh, and in some respects like one that is particularly loosely connected.

Africans can eliminate many or most current restrictions if they establish a large, scalable, and functioning continent-wide infrastructure for data, money, and commerce. They can build a network that technically and commercially links African business and technology communities, while also connecting the African ecosystem with the global one.

Using a political process to connect the people and businesses of 55 nations would continue to be slow and inefficient. Instead, the solution needs to be self-organizing and mathematically governed, and motivated by economic incentives. The right starting number and kind of nodes need to be picked, and then efficiently connected.

A number of 250, maybe even 500 nodes, tightly connected, should be sufficiently secure and fast, and therefore scale out to enable commercial interactions on a continental and even global level. They can be Africa's contribution to the Internet of Brains, consisting of skilled African minds integrated into global digitized supply chains. With its own currency, tied to another Blockchain like an ultrafast BCH, EOS, maybe Stellar or Cardano (natively or using "atomic swaps"),[140] or linked to the still much slower main chain of BTC or

ETH, this then binds IoB transactions of the global digitized supply chains to an effective and globally accepted financial infrastructure.

In its ultimate vision, this network of 250 to 500 nodes would originate in competing but overall complementary innovation hubs distributed across the metro areas of the African continent. As such they would represent a community-based and networked platform with its own commercial and high-tech computing and communications infrastructure. They would have economic incentives for investments and education, empowering the millions of young and skilled Africans associated with them, offering services to global businesses and consumers. Ideally, the platform itself would be scaling out globally, but Africa seems like a near-perfect breeding ground.

This is the general technical network "science" logic. Next, we need to think more about the practicalities of execution.

Translation: Bringing It Together in Real Life

By addressing the major obstacles to Africa's path to wealth, I have previously outlined key conditions for a Blockchain-enabled solution to boost African high-tech and data-based business activities. These conditions can guide our current and prospective heroes of Africa when executing that vision. In summary, they are as follows.

Above all, they start with a focus on the positive and the opportunities. These are the *brains to tap into* instead of *mouths to feed*. A realistic business case should indicate substantial benefits and large ROIs, and a specific plan for how to get there. As our heroes then implement the plan step by step, their vision will seem ever more achievable. FOMO can replace the sarcasm that dominates way too many perspectives on Africa.

Because they are the by far most valuable resource of Africa, the plan needs to almost single-mindedly focus on empowering and enabling young and skilled African brains. For the ever more digitized and data-driven world of the near future, this means building a communication and computing infrastructure that targets almost solely trained African brains. This needs to be profitable and privately funded. It makes the efforts self-sustaining since they become driven by economics. Investors like getting money back. Once they see that it is likely, they will open their checkbooks.

Key to the success of such activities is the pursuit of blue ocean initiatives. Instead of directly competing with existing business models, institutions, and interests, this is additive. It generates new value and revenues out of future-oriented activities. One of the most important conditions here is to reach far beyond the narrow confines and small size of individual African nations, or even of the whole African continent.

For this to happen the value generated must be for the 97% of the global economy outside Africa, and paid by these non-African customers.[141] It is them and not the in comparison poor Africans who have the means to fairly compensate for value-adding activities. This alone will almost instantaneously change the business case for investments and for commercial activities.

Reaching out beyond traditional limitations in a fragmented world requires establishing a continent-wide network of hubs that focus on where the African brains already live. These are not defined by borders of nations, ethnicity, or law. Rather, they are the metropolitan areas, and even more specifically such locations with innovation hubs and educational institutions. The goal is to go the path of least resistance and maximum benefit, reaching for the lowest hanging fruits – and focusing on creating high value via participation in a future-oriented economy.

American criminal Willie Sutton was once asked why he robbed banks. He is credited with saying, "because that's where the money is!"[142] One should focus one's activities where we can expect the biggest results. Our ambitions are nobler than Willie Sutton's, but we should follow a similar rule. The focus of investments needs to be where the biggest assets are – the big cities.

Blockchains Boosting Africa

Blockchain-based technologies and economic models can, and likely need to, be the enablers of this. They can provide a platform to account for value delivered. This enables the bridging of national borders but also becomes the "currency" that links foreign customers to African experts delivering high-value services and products. In the form of a coin, or token, it then also is used to compensate for investments into infrastructure for communication, computing, and training. Arguably, only mathematically enforced trustless Blockchains can introduce a level of governance and stability that instills confidence into such new kinds of solutions. Alternatively, which one of 55 political and legal systems would one rather put one's trust in?

Effective governance includes a wide range of stakeholders based on ownership of tokens, which again is driven by the tangible economic value generated. Among these stakeholders can also be governments and public institutions, although only as in-kind or financial investors. Many of the initial efforts of African-brain-focused solutions would be complementary to governmental functions. In the long term, these relationships may even switch upside down. New Blockchain-enabled approaches for health care, education, or energy may deliver the real benefits. Governmental programs may become complementary to the newly established future-oriented markets and solutions.

The proposed Blockchain-based decentralized model for investments and commerce offers the first continent-wide effective platform for private industry investments. It reduces the associated risks for investors due to the diversity and extent of the platform. The investments facilitate and enable commercial activities that directly benefit Africans who provide high-value services to the rest of the world. This reaches into financial matters also.

The combined expertise and money of rich-world investors, technology organizations, the African Diaspora, foundations and universities will be available to pursue profits and revenues that directly (also) benefit Africans. All this can be leveraged for subsequent activities and initiatives on a continent-wide, national, or even local level, as Africans deem necessary.

Perhaps focusing on metro areas and brains will introduce new disparities and widen the gap to the vast majority of all Africans. This is probably a fair point, at least initially, but with limited meaning. One portion of the economy will continue to grow the low levels of income at a minimum of the same 3% to 5% as today. Another one, however, will almost instantaneously generate per capital incomes multiple times as big.

The alternative cannot be *not* to take an action that injects vast amounts of money, expertise, and wealth into the continent. Rather, the underlying economic model needs to be designed in a way that it involves as many stakeholders as possible, motivates a continuous expansion of the hubs, and spreads wealth across an ever-larger portion of the population.

Infrastructure can thus be expanded into the vastness of the African continent. Instead of taxing business activities, their profits may even be directly redirected to investments that further expand on needs of the wider African populations. Profitable new blue ocean-based business activities can function as investment vehicles themselves. This may even open the door to some newer thoughts in economics, like Glen Weyl's concepts of "radical markets" which consider the elimination of (some) private property and motivate productive activities with more equitable benefits for people in societies.[143]

After all, what do government use taxes for anyway? Besides paying for safety and security, meaning the military, the legal system, and the police, almost all other payments would either be for infrastructure, or for distributive efforts. If the private sector can use motivations based on economic self-interest to expand such infrastructure and its wealth creation limits the need for distributive efforts, this creates a positive cycle.

Investments into infrastructure and wealth building of any kind, as long as they benefit an increasing number of people, should be encouraged and not directly taxed in the traditional sense (e.g. via income taxes). Such investments lay the foundation for a strong future, reduce budget outlays, and overall strengthen the wealth and the resilience of society.

With a goal to expand the number of hubs to about 250 or even 500, the whole continent and all major population centers would already be covered. Spinoffs and startups can take it from there, while governments may continue to focus their efforts on keeping the result socially acceptable.

The roots of Africa's challenges are economic, and so are the solutions.

Act III

Breaking Through: From *Survive* to *Thrive*

* * *

Chapter 15

Dimensions: The 123 of ABC

"VIBRANIUM IS ALL AROUND YOU," said Wakanda's "Chief Scientist" Shuri in the movie *Black Panther* to explain how her African nation could build a city so wealthy and technically advanced that it looked miraculous to a visitor from the United States.

It is a strange statement, though, because it implies that Africans need a miracle to join or lead the modern world. But that is nonsense. It does not take some magical substance from outer space to succeed in a technology-driven world. One does not need a king either. In the opposite, both are toxic, as we could see ourselves in the movie, and as history has proven so often, not only but also in Africa. Valuable inanimate substances and absolute rulers are a bad mix. Somebody takes over a kingdom (or a data center, or everybody's data), and a centralized society is ruined.

Find gold and diamonds, and the wars begin.

As the movie shows, vibranium did not do much by itself. It did not build the city, but people did, using their creativity, minds, and efforts.

Even with powerful, mind-boggling stuff from other worlds, technologies and solutions still would need to be envisioned, designed, and built. It still takes brains, a vision, a good plan, and smart execution. But if you have the brains anyway, then you don't need vibranium, or gold, or oil, or diamonds. Grey matter beats all other matter.

And so, I come back to a statement I made before: Africa's strength and wealth do not lie in its forests, plains, or in the ground. Its future will be shaped by the brains and minds of its people.

"Brains Are All Around You" sounds barely less ridiculous, but it is a better explanation for how Africa can gain wealth and technological advancement. So, using the outline from the previous chapter, let's look at the precise steps required to reach this success.

The execution opens the future to the people on the continent and beyond. This future then is theirs to take. Once Africans are an integral part of the high-tech based global economy of the future, they have no limits.

Of Stars and Gaps

Something fascinating and majestic happens when we look at a clear sky as night falls. Without distraction by artificial light, everything gets darker until we see the first faint sparkles of stars. We only can see the brightest, spread over the wide expanse above us.

But their appearance gives us a sense of the vast dimensions up above us. As daylight recedes, everything gets darker, but ever more stars become visible. Soon we can see a countless number, and the bright band of our Milky Way Galaxy spans from horizon to horizon. There is not a single area not covered by celestial objects of various sizes.

This is what the early-stage technology and entrepreneurial leaders must remember, the heroes of Africa. The beginnings are faint. But they lead the way, with a countless number of followers ready to join the fray. Before they do so, however, they need to be able to visualize the future, even if it seems dimensions away.

For that reason, the seeds of Africa's rapid economic development are two-dimensional. Boost centers thinly spread across the continent defines the vertical dimension, directly targeting an ambitious vision of the future. They are the first sparkles indicating and achieving the rapid economic growth and opportunities that come with a focus on emerging technologies. Over time they get stronger and bigger, and they may stay the brightest. But by themselves, they are just a beginning.

Although the vertical step has to come first, soon after the second dimension can start. Technological empowerment and liberalization then need to spread out across the horizon and fill the wide gaps that make up the overwhelming number of Africans. New initiatives across the rest of the continent reflect the myriad of personal visions and entrepreneurial ambitions of hundreds of millions of minds and brains that need to join in to achieve the goal of radically advancing African societies.

Both dimensions are complementary, with the first being more critical in the short-term, and the second deciding the ultimate success for everybody.

Vertical and Horizontal Dimensions

The vertical dimension is not geared toward the masses of people but has one single-minded and narrow focus: rapidly enable the contribution of smart African brains to the emerging digitized world of tomorrow. It maximizes know-how transfer and revenues that then can be put to further use in Africa. Because of this focus, the vertical portion of the plan is likely going to directly touch a number of people counted by the mere millions.

It concentrates on hundreds of what I call "boost centers" to promote and enable the selling of services to the rich world. These services are profitable blue ocean initiatives that appeal to private investors. Therefore, they need to show great profit margins, and be distributed across the continent but linked to each other via decentralized Blockchain governance. Focusing on Africa's competitive advantage of large numbers of skilled young brains, together they scale out across Africa, or even globally. This brings massive amounts of income, a significant influx of money.

As a vertical platform, the network of boost centers delivers high-value activities. They originate in all the major African business hubs, typically are in large metro areas. Although only a small portion of the population will be directly involved, not more than between 1% and 5%, the effects will be felt everywhere.

The big impact of this step is that it establishes a functioning platform for commerce, creates a state-of-the-art communication and computing infrastructure, sets up an avenue for continuing investments, facilitates intellectual know-how transfer, and empowers millions of people spread out across most or all major population and business centers on the continent.

Above all this approach is straightforward to use, scalable, and its decentralized structure lowers risk. It injects Africa with private capital *twice*: first through investments, and then through the selling of high-value services and products.

The second and horizontal dimension is as important, if not more so because its long-term repercussions are much bigger. It massively scales out transformative infrastructure. This delivers to the masses of the people what they need to survive: education, communication, health care, energy, and transportation. Once these are in place, people can then put their energy and brainpower into efforts to thrive.

New mechanisms enable a radical transformation of the way whole societies are organized. Without corruption, centralized control, and constraints, this extends the benefits of essential infrastructure across the whole population by applying concepts of the future. Unique token mechanisms combine emerging technologies with new economic motivations. They reduce risks, increase resilience, and establish profitability. This makes private investment possible.

It extends the equivalent of boost centers to all people. Like the vertical dimension, the horizontal one is also based on decentralized private and community-based initiatives that re-organize traditional functions of societies in novel ways.

Today's typical approaches are based on delivering infrastructure via top-down planning. They tax people and then channel their money through a centralized and often politically controlled bureaucracy. By liberating ourselves from the intellectual shackles and constrained vision of that old world, new technologies and decentralized economic motivations can prove to be not just more natural and organic, but also more efficient and effective.

The first dimension will become a key motivator, enabler, and driver for the second, although both can be pursued and start simultaneously. In the long term, the horizontal dimension will become even more important than the vertical one. It delivers the empowerment and enabling of not just several million, but of over one billion people, particularly the 800 million young African brains that I keep talking about.

Infrastructure 1+5: Empowering, Not Telling.

Empowering people requires trusting them, learning from them, and to be open to enjoy the fruits of their creativity, even if all of that happens organically, "chaotically," and with lots of uncertainty. Therefore, I will not speculate about how exactly African liberated brains will use their new powers. They are set free to pursue their own dreams of happiness. This is the part of the story they must write themselves. It is about specifics of *their* future.

Any use of new technology can improve Africa's competitiveness. Most of it is incremental, others more transformative. It all matters, is laudable and should be pursued. What we are here primarily after, though, is giving priority to aspects and projects with genuinely *transformative* or even *transcending* impact. They are about leapfrogging, directly into the future.

These are related to infrastructure, and they directly empower and propel young and skilled African minds and brains. It requires six aspects of infrastructure. I call them the "1+5." On the vertical dimension is a *financial* platform for exchanging money, and its *commercial* counterpart that links skilled brains to each other and to investors. Smart-contract based, and using a coin, it defines and executes the myriad of commercial transactions of the IoT and IoB. This platform is not dependent on foreign powers or banking interests, or on local governments and other intermediaries. And it still delivers a hard and value-based currency, money, and a way of making more money.

The five horizontal infrastructure aspects are crucial for any African's ability to contribute and create value: *communication and computing*, *education*, *health care and biotech*, *energy*, and *transportation*. Infrastructure does not create value but is a necessary condition for value to be generated. Blockchain-based economic motivations and rewards become the main drivers and conditions for vertical and horizontal infrastructure development because they can use tokens to link the benefits of future value to today's investments. Investors can then invest into infrastructure because they benefit from the value that this infrastructure creates.

Both dimensions have in common that they bypass existing established interests. In that, and if governed mathematically and in decentralized form, they transcend traditional issues of corruption, control, and constraints. They are blue oceans and enable genuine leapfrogging. This makes them additive, meaning they can co-exist with other solutions in place. Due to their innovative approach, they even support them. If they succeed, society overall wins. They will then eventually supplant the old ways. Energy grids will remain localized and decentralized, health care will be based on personalized genomics and prevent instead of dealing with the sick, and education and data connection will stay linked to their practical application for society.

The five areas of horizontal infrastructure should be prioritized based on how much they develop, empower, and enable African minds and brains. The closer they are to the brains, the more they directly empower them, the more value they contribute on an economic and a personal level. This defines the priorities. First, is education. Second, comes communications and computing, then health, followed by energy and transportation.

Africa Boost Centers

Technical "boost centers," or Africa Boost Centers (ABCs), is the term I use for the many organizations connecting the African brains to the high-tech global economy of the near future. Ideally, they are rooted in innovation hubs. By being the physical location and financial and commercial conduit, they turn into the nodes (in technical terms the "crypto-technological consensus validators and bodies of governance") that link infrastructure investments to educated Africans, and skilled African brains to their high-end markets.

Together, the ABCs make up a decentralized network, bringing them out of the control of political entities. They are governed by extraterritorial or even totally encoded bodies, and by economic rules inherent in a coin mechanism. A high number of ABCs mitigates investment risks because they compete and "win" via efficiency, among each other and compared to the rest of the world. Risks will overall average out. Even maliciously motivated physical control over some centers could not achieve much. Work would simply be channeled to other ABCs.

As nodes in the overall logical, physical, and commercial network they will have secure facilities, high-speed Internet connections to the outside world, office space, a localized 5G infrastructure, and profound (edge) computing power. ABCs align with training institutions, tech companies or foundations, and VCs, the African Diaspora, or other sources of investment.

They operate and leverage a joined coin, used to coordinate investments and the commercial platform, the Internet connection, and the link to education and training.

A 3-Step Boost

The network will ideally be built in three steps. Step 1 focuses on no more than 5 to 10 ABCs, whose roughly estimated overall turnkey cost are in the dimension of $5 million to $10 million per node.[144] Steps 2 and 3 will expand the numbers of nodes by factors of 5 and 10. While this happens, the numbers of brains delivering high-value services inside each ABC keep increasing also, from initially just dozens and hundreds of people.

Investments of an estimated about $8 million to $15 million should be sufficient to kick off the project. Using these funds can build the basic platform, a prototype, and a turnkey ABC reference model. It also includes the signing up of ABCs, defines the exact business case and technology components, establishes the stakeholder community, and gets investment commitments. This bootstraps and kickstarts the network.[145] Each boost center also develops expertise in at a minimum three of the five horizontal investment dimensions: communication, education, and health. All three directly benefit the women and men working through them and are therefore critical to the centers' success.

A preparation stage also includes the coding and development of the Blockchain-based financial and commercial platform and establishes the corresponding governance principles, and economic reward and penalty mechanism. About 1 to 3 sites will need to be involved.

Then the network can be built out.

1. **Pilot.** With investments of between $25 million to $50 million, about 5 to 10 ABCs can be established. The locations should be spread out across Africa. A likely focus could be on East Africa (e.g. Nairobi, Addis Ababa, Kigali), but it also should include one West African (e.g. Accra) and one Southern African tech hub (e.g. Lusaka).
2. **Launch.** Using investments of between $125 million and $250 million, the solution can be scaled out Africa-wide. About 25 to 50 ABCs can be brought online.
3. **Roll-out.** With an additional $1.25 billion to $2.5 billion, the solution can be rolled out until it reaches a pre-determined ABC limit of 250 to 500 locations. The later expansion would be through satellite ABCs.

These three steps mitigate risks while rapidly achieving the benefits outlined before. If the activities are executed in a decentralized crypto-controlled community of stakeholders that include VCs, technology foundations and companies, the African Diaspora, and rounded up by large businesses, governmental, and supranational institutions, they provide an excellent avenue to invest, with outstanding ROI potential.

The Real Fun Can Start Then: Letting the Brains Take Off

The skills, money, and financial and commercial infrastructure of the ABC network establish an environment conducive for startups and spinoffs. The more the brains learn how to prosper using emerging technologies and approaches, the more they will be driven toward entrepreneurial activities that address daily challenges in Africa and the world. This will trigger the emergence of a technically oriented entrepreneurial ecosystem surrounding each ABC, permeating the African continent.

Prospective heroes of Africa's future turn into actual ones, and their numbers expand rapidly. They are the brains unleashed.

Now the real fun can commence. There is no inherent limitation anymore. Distances and borders have been overcome. Independent and decentralized platforms exist. Basic needs are being taken care of without having to rely on political processes. Governments are no longer restrictive but concentrate on their core functions, not the least safety and helping to organize social priorities. Investments can effectively be channeled into a variety of creative and commercially viable solutions.

This is a rosy picture. But first, there is more work to be done. The six infrastructure components need to be addressed more specifically.

Since money is the main condition for success, let's start with having a more detailed look at the vertical platform that is supposed to deliver exactly that.

Chapter 16

There It Is: Money for Africa

GETTING MONEY to Africa would be easy – if it just were easy! And fast, cheap, and reliable also. Today it is not.

I was standing in the printer shop near downtown in Kigali with a credit card in my hands. Although the walk from my hotel was all downhill, I had started to sweat. Inside the store it was even warmer, with dozens of people working busily in a clean but cramped space. Business seemed good. But although I was willing to do so on the spot, I was unable to pay, and the merchant could not take my money. The ATM cash machine could not dispense as much cash as I would have needed to settle the bill, I did not have mobile money on my phone, and the printer did not take my credit card. A couple of weeks later, after filling out a bunch of paperwork and paying multiple kinds of fees, Francois finally received the money that he had well earned by doing a quality job.

Before the world can really do business with 800 million young African brains, or over one billion people altogether, it needs an effective financial system that supports such commercial activities. Looking at today's situation, though, it would need to deal with 55 weak national currencies that nobody wants, plus US Dollars or Euros, multiple layers of transactions fees, exchange rate fluctuations, currency controls, laws and regulations subject to political instability and influence, and lots of time and nerves. By the time one has tried to make sense of it, business has moved on.

And so, the world keeps losing out on over one-fifth of its most ambitious and creative brains, at a time when it can afford to do so least. And Africa continues to miss out on the huge revenue potential of selling these brain's capabilities and their contributions to improve the world. In an ugly twist of fate, the entrepreneurial technology community of Africa is largely excluded from doing what it can do best: use technology to come up with creative solutions for practical problems.

Commerce requires a stable currency to calculate prices and profits. It also needs the ability to effectively define and track agreements between two or more parties to determine for when money is to flow, and efficient tools to use when paying. Business is limited without such an effective and trustworthy financial and commercial infrastructure.

This has far-reaching effects. African entrepreneurs are full of ideas and initiatives, but their ability to succeed on a global scale is low without money and without the infrastructure that money can afford. They are constrained by circumstances that force them to "survive" instead of "thrive."

Overall Africa is the continent with the lowest bank branch penetration,[146] although there are big differences between its countries. Telecom operators make up for that to a degree. Their vast networks enable them to easily reach customers on a massive scale.[147] Banking in Africa is not just growing fast but it is at the same time also hugely profitable,[148] more so than in almost any other global market. Clearly, banks and other kinds of financial intermediaries know how to make people pay for using their services.

But commerce, finance, and banking in Africa should not be difficult or costly. "All that is needed" is combining the ease of use of mPesa, the stability of the US Dollar, and the limitless capabilities of a cryptocurrency, in a way that motivates investments into future-oriented skills and technologies.

Once the outside world sees an effective way to integrate skilled African brains, or even products, into their digitized supply chains, private investment money will flow. Everybody benefits from 800 million young and flexible African brains unleashed.

Backing Up: The Challenges

The main requirement for such a financial platform is to create a *victorious* cycle that breaks the current vicious one. It requires skills and infrastructure for Africans to make valuable contributions to their own economy and to the global one. Developing such skills and infrastructure needs money. But without making valuable contributions in the first place they cannot be compensated, and money does not flow to fund such important investments. Africa is then mostly left with governments slowly extracting some investable capital by taxing small economies and a generally poor population, or staying dependent on gifts, donations, or investments sponsored by foreign nations, of course with strings attached.

This vicious cycle says *no money no output, no output no money.* Forget about advancing rapidly. Even though it should be the other way around, in reality, the name of the game is *survive, not thrive.*

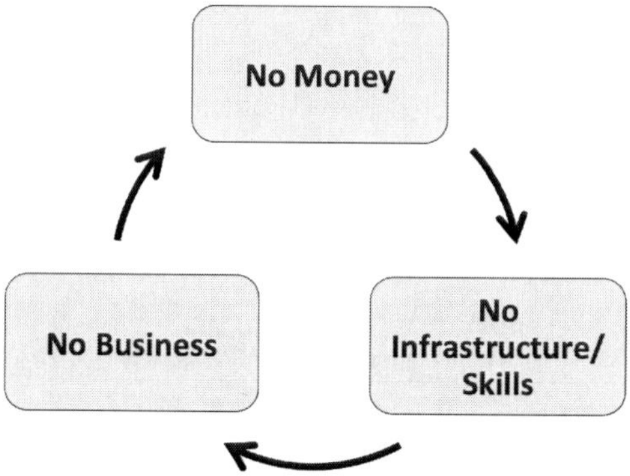

Figure 7.
"Survive, not Thrive:" No Money No Output, No Output No Money

The culprit here is that as desirable as the contribution of African brains is, as difficult it is to invest. Money cannot easily, cheaply and reliably be sent back and forth, which is one of the most basic requirements for any commercial activity. Where to begin when each one of 55 nations has unique pressures and priorities?

Today, Africa has no stable, affordable, and universally accessible means of performing financial transactions with other nations, whether these are different African countries or on other continents. Transaction fees and exchange rates are exorbitant, often 15% to 20%, and sometimes more. On top of these costs, paying from abroad is also difficult and slow, and exchange rate volatility is high.

Wires in local African currencies are not feasible, or impossible at all. Rather, they will need to be made in Euros or Dollars and will then be translated into local currency when received. Payments of invoices denoted in local currency are difficult, essentially requiring the setting up and maintenance of a US Dollar or Euro bank account, and related accounting. This does not make sense for most businesses.

The overall result is a stifling of cross-border business activities. While this may not be a large issue between Malawi and Mali, not being able to efficiently sell products or services from Nigeria or Ghana to the United States or Europe is more likely to be a big deal.

In a meeting with an East African central banker in May 2018, he described to me how even regimes like the East African Monetary Union (EAMU) are rather inefficient. It still is the case that the U.S. Dollar is the de facto means of exchange between East African currencies. The only realistic alternatives would be basing such transfers on the Euro or, just maybe, on the South African Rand. Each requires exchanging currency twice and incurring related fees and delays. And it still leaves African economies dependent on fiscal and monetary policies of foreign economies. Such economies, or their political leaders, are highly unlikely to consider the interests of small African economies in their decisions.

Limitations go beyond payments for cross-border transactions. The same inefficiencies apply to investments, including venture capital, and to savings. Not only are there fewer investments and savings when transferring money is inefficient and costly. The impact of large exchange rate fluctuations is also compounded by the longer time horizon associated with them. Cross-border crowd financing or micro-investments are difficult to manage if there is no stable currency.

In such an environment, try raising substantial investments from international sources for infrastructure projects – in an African local currency. It just will not be easy, if it is possible at all. Current national African currencies cannot do it reliably. Something more stable is required.

The potential of current cryptocurrencies is also still limited. Eventually, some or all of the current performance, usability, and cost issues may get resolved. But today the big coins like Bitcoin or Ethereum are not there yet, and above all their high exchange rate fluctuations are still too impractical for daily cash transactions. Others like Ripple, EOS, Stellar, and Cardano do on top of that not yet have enough global or even regional market penetration.

With so many Africans depending on cash at hand, most people simply cannot afford not knowing how much food or clothes they can buy with a crypto-coin when exchange rates between the crypto-coin and fiat may continue to change by 5% in a day, 20% in a week, or 50% in a month.

Without a capable financial infrastructure, Africans cannot compete or contribute on an equal footing with the rest of the world. How can they thrive if they start with such a massive disadvantage? The high cost of capital creates dependencies on NGOs, foreign governments, or multi-national institutions. But African entrepreneurs don't need handouts. They can contribute and compete, as long as they are given a fairly level playing field. Then they will forge their own future.

Despite all their shortcomings, Blockchains can play a major role in making this happen.

Why Blockchain Can Help

There are two major sources of money for Africa: foreign markets and new technologies. Both are potentially massive. Non-African markets still comprise 97% of the global economy. Today, Africans are but an afterthought in them. Everybody wants access to rich economies, but most other countries are better connected and already integrated. Since the world's money is outside Africa, though, establishing an effective infrastructure to do business with the world is not an option, but a must. That infrastructure needs to support financial and commercial activities and has to be based on high-speed Internet connections.

So, is all lost anyway?

Not if we consider the second source of money, which is a blue ocean. It is the term describing a market for valuable new services, products, or solutions that did previously not yet exist. In Africa's case, those are based on emerging technologies like IoT, AI, genomics, or AR/XR and require mostly one additional ingredient besides the very same commercial and data infrastructure just mentioned. The extra element is trained human minds – skilled brains.

The combination of both is where the real value lies, selling the capabilities of skilled brains into foreign markets. It is the sweet spot where Africa's technology ventures gain the biggest results for their investments.

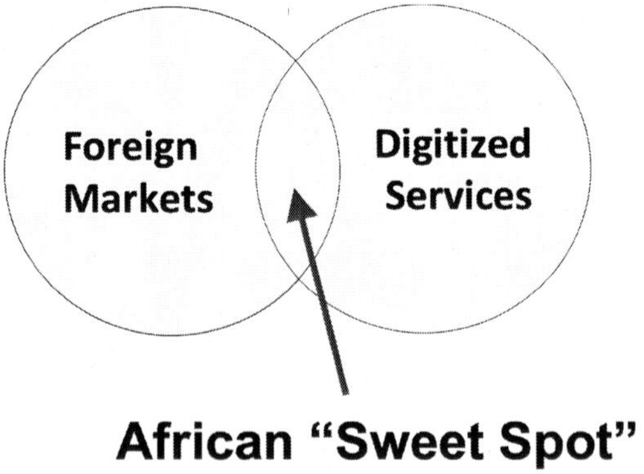

Figure 8.
Africa's "Sweet Spot" for Making Money

Despite their above-mentioned shortcomings, Blockchain technologies are still the best tools for this challenge, and natural ones at that. Implemented effectively, they are borderless and decentralized, efficient and scalable, and they can include flexible economic motivations that facilitate and encourage commerce and investment.

According to its original White Paper, Bitcoin as the world's first widespread Blockchain currency is "A Peer-to-Peer Electronic Cash System."[149] Fintech can therefore be the heart of what Blockchain can enable. With fiat currencies, people all over the world have already agreed on using tokens to keep account of value and the exchange of value. This is what happens when I give you a piece of paper in form of a banknote, or a bank changes a financial record in a database to reflect me paying for a product or service. We are all familiar with this. Concepts of Blockchain then just add an additional aspect to it, although a powerful one.

When a currency is peer-to-peer there are no middlemen and no borders. Efficiently implemented, the implication of such a concept is very profound and powerful. It also directly addresses many challenges that are particularly prevalent in Africa.

Imagine the overall impact on the African economy if there were no financial borders and no need for banks, and if money could be invested or sent flexibly to, from, and across the whole continent.

On top of all of that, Blockchain currencies have already proven that they can solve a lot of the same challenges of African societies that I just outlined. Despite all their limitations, cryptocurrencies have by now reached a combined market value of hundreds of billions of US dollars. Bitcoin and Ethereum are used for commercial transactions around the globe, and other crypto-solutions have even outdone them by improving performance and lowering costs (i.e., EOS, Stellar, Bitcoin Cash, possibly Cardano).

Equally important, cryptocurrencies raised billions of dollars of investment capital to be used for their own development of coin-based solutions. Similar tools may help in funding other tech or business ventures. Nothing makes raising capital via cryptocurrencies limited to building such currencies in the first place. Rather, it may be one of the worst reasons to invest, compared to funding health care, a decentralized energy or communications infrastructure, or education.

Blockchains function by mathematically establishing economic motivations, using tokens that represent either existing value, or more importantly value to be generated by using the token. The same principles can also raise capital for building the infrastructure required to create and exploit this value in the first place. This exact mechanism can be leveraged for Africa's biggest assets, which is the potential output of their young, creative, and ambitious brains.

A pan-African or even global cryptocurrency can help unleash African brains in dramatic form. Blockchain mechanisms can create a victorious cycle by motivating continuous investment into a realistic and profitable version of the future. Instead of "no money no output, and no output no money," a token can link future value to today's investment, and thus encourage such investment.

It then means *money, therefore output, and output, therefore money.*

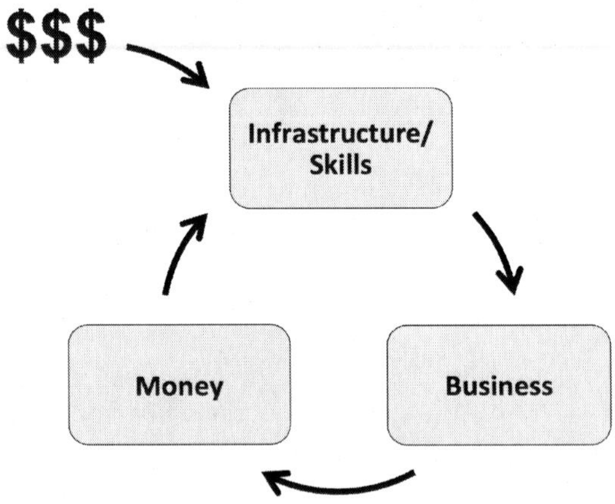

Figure 9.
"Thrive, not Survive:" Money, Therefore Output, Output Therefore Money

A Decentralized and Borderless Crypto-Platform and Coin

No matter what, the African financial infrastructure needs to be decentralized, controlled mathematically and outside political influence, and be borderless. It needs to be usable across the African continent, and beyond. The opposite approach, purely creating a national Blockchain-based currency, would solve none of today's profound issues, as long as it was subjected to the same or similar economic and political limitations as the current fiat currencies.[150] In effect, it would likely be just another fiat currency, although using a different tool.[151] The technical aspects would simply be "tinkering at the edges."

This begs the question of why not simply use one of those borderless currencies that already exist, like Bitcoin (BTC), Bitcoin Cash (BCH), Ethereum (ETH), EOS, Ripple (XRP), Cardano (ADA), Stellar (XLM), or others? This is what Stellar, Bitcoin Cash, and Cardano were advocating to do, during our TAS2018 Blockchain events.

The answer is that in principle they should be considered since either of them could in theory do the job. Even pegging a new coin to one or several of these currencies may be a high-value option.

If we stopped here, we would already have achieved a lot, essentially economic freedom based on the ability to use cryptocurrencies Africa-wide for commercial transactions. The biggest limitation would then be that these currencies are generic and do in their current form not contain such additional specific and substantial economic motivations that we are looking for.[152] If we generated investments in education and infrastructure to connect and integrate African brains into a future-oriented economy, we would need to bolt on such motivations to an existing currency, or create our own coin.[153]

At some level it would in principle not even matter whether the economic model is based on a permissioned or a non-permissioned platform, meaning whether it is open for the public to join or requires some form of "membership."[154] My next statement may be anathema to the global Blockchain community, but the new currency could even be loosely modeled on the American Federal Bank. The Fed is a private institution established by private businesses, in this case, banks, in which the federal government plays only a limited supporting role.

Instead of using such a cartel of private banks and politically or business-driven processes, our new currency would hopefully contain different motivations and governance principles. These need to be linked to the value to be optimized, which brings us to the second piece of the puzzle.

Identify the Core Value that the new Coin is Tied to

To motivate the creation of blue ocean solutions and revenues, the token's value needs to be directly linked to such value. This cannot be coffee or diamonds since in the big scheme of things these are just small blips on the world's economic radar. More generically, Africans should ask what the future world will look like, how they can best contribute to it, and what they need for it. This then is value that is worth investing in. It, therefore, is critical to ensure that any cryptocurrency is linked to such future value. For my example, this would be African brains, or rather the output of African brains, or at a minimum the ability to create such output.

I have already laid out that skills-based education and high-speed Internet connectivity is an absolute must for Africa. Such investments must be economically motivated. A substantial portion of the coin's genesis block should, therefore, be linked to them.[155] This is not sufficient, though, since infrastructure alone is not "value" in itself. It just *enables* value creation.

Here are some additional pointers on how to identify future value generation and align and take advantage of a greater community and its network effects.

The world of the near future will largely be based on digitized supply chains, the design and building of solutions leveraging AI, creative design using AR/XR, and integration into manufacturing using virtual services and IoT. This requires high-speed Internet access up to the level of 5G. Above all, it means that the token needs to incentivize investments in providing skills-based training, establishing high-speed Internet connections for trained people, and integrating their remote services into digitized global supply chains.

This is far superior to using an ICO as a generic mechanism for investment. The coin value is linked to tangible real-world value, motivations are more clearly defined, and the mechanism establishes and enforces network effects.

Other options for defining tangible value would follow a similar logic, although mostly on a smaller scale. For example, if one were to build value by motivating the decentralized generation, storage, and use of energy across Africa, then such a process could warrant creating its own coin. Energy generation and storage capacity could be the asset backing up the initial coin, and effectively running and maintaining these capabilities could generate additional value.

Build Economic Motivations that Link the Value to the Coin

The costliest aspect of Internet access can be provided via Low-Earth Orbit (LEO), directly connected to hundreds of technology "boost centers," which originate in innovation hubs across Africa. There, educated Africans get trained. Investors provide capital to build this infrastructure, train African brains, and link them to a high-speed Internet connection, in exchange for a share of their service revenue. That revenue is measured by tokens, and the share is encoded into the solution itself.

The governance and economic consensus mechanism should be centered on the value generated by these trained brains. For them to be most effective, they will need to use 5G, a dedicated high-speed Internet connection, and edge computing capacity. Their use can be directly linked to their providing of services. The mechanisms for that can be determined by Blockchain-governed smart contracts. The initial coin value is backed by investments and increasingly so by a portion of the revenue channeled through this platform.

The coin's main function is to govern the overall value-generation process and to provide access and the right to use the infrastructure. Since it also becomes the internal payment mechanism, its coded and mathematically enforced rules can ensure that the value generated does properly compensate each stakeholder. In other words, no investment no value, no training no value, and no work no value. Everybody's motivations are aligned.

More positively expressed, investments plus training plus work equals lots of value.

The currency is initially distributed to investors and founders via a coin genesis block. Subsequent coins can be earned through the work of individuals, thus replacing traditional processor-based Proof-of-Work (PoW) "mining." New tokens will be created as more trained Africans join these global digitized supply chains and provide services. Everybody benefits, and increasingly so, since new coins are being created as Africans earn them while generating income during their providing of high-value services. The validating nodes are the hubs, or Africa Boost Centers that govern on-chain via some form of delegated proof of stake (DPoS) based mathematical consensus mechanism.[156]

Value will be "internalized," meaning that the economic value creation enabled by infrastructure does in effect back up the currency's price. When people use AR, computing capacity, the Internet infrastructure, and the skills they learned in the innovation centers to remotely work on product design, the revenues they generate will be translated into tokens and shared among all parties. Money flows back to the investors, trainers, and individuals, as defined in smart contracts.

This can be achieved in several ways. In principle, it relates to linking the value of tradable physical or digitized products to the token.

Boost centers based on innovation hubs seem like natural drivers of leapfrogging. The logic is similar as when mobile phones replaced bank accounts after mPesa came about – they are already there. We create value by providing high-speed Internet access to technofriendly locations, training capable minds, and integrating them into global digitized supply chains. Reaching out, training, and empowering individuals is close to what African innovation centers already are doing today.

The new token becomes the mobile phone minute for digital work. A big distinction is just that it is governed by a community and not controlled by a single business.

The Role of Governments and Other Stakeholders

Governments then need to make some bold decisions. Eventually, they have to choose whether they want to unleash their people's brains, ambitions, and economic drive, or whether they fear being challenged by the resulting dynamics.

Besides completely resisting any Blockchain or crypto-technology as means of payment, in principle, this leaves them with several options. Since the Blockchain mechanism that I just outlined is based on a blue ocean approach generating massive value where none existed before, governments can (and should) explicitly deregulate such currencies' use and "let them run." They lose nothing, in the opposite. The success of the above plan means that millions of people will use state-of-the-art infrastructure. They and their families will not require government assistance, in the opposite. Taxes may still be in place, although I would rather prefer that they are limited or abolished, given infrastructure and economy-boosting activities triggered by ABCs.

In the next chapters, I will lay out how societies can leverage Blockchain technologies to fund some of the core services that governments currently provide. Some of this is relevant here, too. Governments can also directly support and facilitate the establishment of the network of African boost centers. This could turn them into stakeholders, with the ability to benefit via coin ownership.

And, via a token or currency backed by the value of high-tech services to rich economies, governments can gain more stability of their own economies without being subjected to fiscal or monetary policies of other nations.

Tick Tock. Mind the Clock.

In American Football there is a "two-minute warning." In European Football there is "stoppage time." Both indicate that the game is drawing to an end, that the outcome will be decided soon. We may not be quite that far yet, but time for Africa is limited to get going with a future-oriented financial infrastructure.

A stable pan-African currency governed by distributed consensus and validation nodes based on boost centers growing out of innovation hubs would create substantial benefits for commerce in Africa. And yet, there is no definite need for an Africa-only coin. Even just using Stellar, EOS, or Bitcoin Cash could potentially achieve a large portion of this effect. Using such existing currencies could even have important advantages since they are at least to a certain level proven to be scalable, and their success depends on them being secure.

A stable currency does not need to support all business activities either, and fully compete with fiat currencies. The above plan outlines the use of such a crypto-coin for a subset of African economies, to help fund and then run new and future-oriented commercial activities. This is not too dissimilar to the global use of the U.S. Dollar for trade in petroleum products.

However, to get more ambitious and take advantage of much more profound, larger, and faster benefits, there is a unique window of opportunity today. Africa better start taking courageous action soon. LEO and VLEO satellites are being sent to space as I write this, digitized supply chains are being established right now, and Africa has a unique chance – to be a founding member of this future world, and to contribute in shaping it.

Africa's biggest risk may be to get stuck and, again, having to play catch-up. By being too timid and focusing on catching up and only taking small steps, African nations may not even get to where they can worry about falling into something as advanced as a middle-income trap. Limiting oneself to building infrastructure the traditional way and pursuing the low-cost labor strategies of the past may cause hitting a wall much earlier. Should that happen, not becoming an integral part of when humanity builds its future would be the biggest failure of the current generation of African leaders.

All it takes is courage, vision, and trusting the positive effects of relinquishing control of a future-oriented initiative (that otherwise would not be pursued anyway) to a large and diverse group of African and global entrepreneurial minds and an increasing number of educated and skilled African brains. A plan as outlined above does not require more money and taxes. It does not establish one-sided dependencies. It is business and private enterprise at its finest.

Shooting for the Stars – And Reaching Them.

Establishing a scalable, stable, decentralized cryptocurrency platform that enables pan-African trade and trade of high-value services with the rest of the world levels the playing field as Africa has not seen in centuries.

It makes the best use of Africa's biggest assets – young, dynamic, hard-working brains. By doing so it adds value and creates new riches where there are none today. The African heroes, whom above I labeled the "entrepreneurial technology community of Africa," can then go all the way.

This community can then independently build and contribute to the future not just of Africa, but of the world.

A Blockchain-enabled financial platform may be the key ingredient to unleash the potentially greatest resource on our planet: 800 million intelligent, eager, hustling, creative and driven minds belonging to Africa's young generation.

Chapter 17

Communication. The Internet of Brains

TO THINK OF AFRICA'S young minds and brains at work in the near future, imagine being one of these 21-year-old young, maybe 5 or 10 years from now. It still is early morning when you arrive in the office building that you are working from, somewhere in an African city. This could be anywhere. You just arrived from home, where you helped your mother open up her store. As you get ready for the day and sit down at a small workspace, you take out your laptop and your VR/XR headset.

Now you need to be able to go online, take a ticket and design that little piece of furniture, that widget, advise a customer somewhere in another place of the world, check the robots, resolve a technical issue, program a 3D printer, perform quality assurance on somebody else's work, or perform any other job in the new digitized world of manufacturing. It is mid-21st-century manufacturing, or as I call it "Digitized Individualized Production" (DIP).[157] [158] No factory lines, but digital work, consumer-focused, personalized by adjusting a specific version of a product to the customer's unique wants and needs.

As you don your XR headset and log on via eye and voice scans,[159] the most important thing you need is an extra high-speed Internet connection. It needs to be reliable, with high throughput and low latency capability. Near-real-time interaction with 5G capabilities is required for many parts of your job. You have to be able to see the specific tool in real-time, or the living space where you are trying to fit the piece of design, and sometimes even sit with the customer in their living room or check out the machine to clarify specs and priorities.

This is part of the communications infrastructure that has to start by focusing on vertical integration into the global digitized supply chains, to then be expanded to everybody across the continent.

Somewhere at a Desk in Africa...

In mid-2018, it is safe to assume that few if any Africans have such Internet connection performance. And still, it arguably is the most critical physical infrastructure component that needs to be in place to unleash African brains and let them contribute as an integral part of the global digitized supply chains.

If you don't have this connection, somebody else will. That somebody could be somewhere else in Africa, or somewhere else in the world. They will sit there with their headset, use the same translation tool when talking to the customer, display a similar Avatar, look at the same virtual widget to be produced or machine to be fixed. Like a couple of decades earlier in the anonymity of the masses physically toiling in the large halls of East Asian manufacturing floors.

This time, though, manufacturing is virtual. Other skills matter, not the dexterity of the fingers when physically assembling a product, but brains. Africa has more of those than anybody else.

But it also needs high-speed low-latency Internet connectivity just like you would be sitting in the same room with your customer. This simply is not optional for Africa. It is a condition to prevent staying an afterthought in the modern world, relegated to the second tier.

Establishing high-speed Internet access is the hardware condition for the full empowerment and unleashing of African brains, particularly the 800 million young ones. Without virtually instantaneous response times while transmitting data on a massive scale, the integration into high-end digitized supply chains is not realistic. Smart algorithms can help, but no matter what, the digital experience and interaction speed needs to be "real."

A high-speed Internet connection with near-real-time performance. It is transformative almost by just being there.

So how do you enable digital workers doing their job using high-performing Internet at this ominous location, "somewhere at a desk in Africa?!"

Challenges Made for Blockchain

We have to solve three main challenges. First, we need a suitable data link to and between the major African commercial centers. Second, we have to connect the end users. Third, we then need an automated commercial platform to pay for data access. The first requires swarms of low Earth orbit satellite swarms. The second needs 5G relays and edge computing capabilities. The third one needs a creative economic model and an IoT approach to using data. And building all these three infrastructure components needs to be funded.

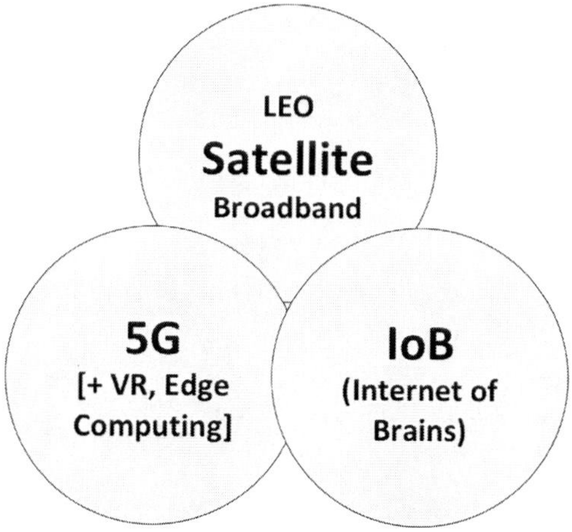

Figure 10.
Three Components of a Future-Oriented African Communications Infrastructure

In all three of these, Blockchain can shine big-time.[160]

Blockchain can make the accounting and paying for data transfers more efficient. That by itself is important but not really a big deal. Its unique contribution is that it can also introduce profoundly new business models to enable the payment for Internet connectivity, and drive the required infrastructure investments.

The fun aspect here is that it puts a new twist on what the crypto-community prides itself for: empowering individuals and bridging borders. Here, though, the goal is funding future-oriented investments that integrate African brains into the economy of the future.

Using new crypto-economic motivations and rewards, it does nothing more and less than leverage classical economic motivations. And suddenly we have – high-speed Internet connectivity for millions of productive Africans!

Let us walk through this step by step.

Step One: The Mother of All Pipes

The first piece of the puzzle is physically providing high-speed Internet connection *into* Africa. The data connection needs to go into the continent and between the major African centers of business activity. Lagos, Accra, and Addis Ababa have to be connected as fast to Tokyo, San Francisco, and Berlin as Tokyo to San Francisco, San Francisco to London, Berlin to Shanghai.

Traditionally this would happen via big fiber-optic pipes and mobile phone towers. Many African countries used enough of them to leapfrog traditional landlines.

This is not sufficient because Africa's data speed and throughput connecting to the Internet backbone is not big enough.[161] Things need to scale out massively. But luckily the Mother of all Pipes (MOAP) is right above us. And so, we go to space. Not us, since already at least half a dozen organizations are doing it. Medium Earth Orbit (MEO), Low Earth Orbit (LEO), even Very Low Earth Orbit (VLEO) micro satellites and nano satellites can address the needs.

Building new fiber optic pipes to connect people may play a role, but in reality it is "so yesterday." Satellites are the answer. Not the big 22,000 miles above Earth geosynchronous ones in High Earth Orbit (HEO). Those play a role for what is called data backhaul but the noticeable time delay leaves them with limited suitability for 5G-style applications. Due to limits of physics, their latency will always be too noticeable to be considered "instantaneous."[162]

Large numbers of LEO micro-satellites anywhere from 100 to 1,200 miles above the Earth' surface change the rules of the game. They cost relatively little and provide improved support and maintenance options. Chances are that when SpaceX sends up a big satellite there is lots of room to pack in dozens of smaller ones also that can spread out and cover large parts of the Earth. Including whole, large continents.

Like Africa.

Iridium was one of the first businesses articulating and pursuing this vision, although initially with mixed success. It is far from the only provider, though. Intelsat establishes powerful capabilities in MEO locations. SpaceX' Starlink, Teledesic, Globalstar, OneWeb, Telesat LEO all work on "high-throughput low latency" satellite constellations. Many have launched satellites. By 2022 Africa should have the choice between four or five providers. Increasingly more of them will be HTS, high-throughput satellites that can deliver bandwidth in the Gigabit per Second (Gbps) range. While that is still not enough to cover the majority of Africans, it is enough to run 5G applications for limited numbers of users. Hundreds of small satellites can connect millions.

Now we need to look at the next step, figuring out the business case and how to pay for these services.

We get closer to that when looking at Step Two.

Step Two: Connecting (Commercial) End Users

The next question to ask ourselves is who these end-users are that we need to connect. And sure, there is no immediate need (yet) to connect the average African consumer to the high-speed data connections mentioned above. They would not typically need to buy these services. African's customers in the rich world do, from outside the continent.

These customers can pay to enable African brains to contribute to their work. They may be consumers themselves or more likely businesses participating in the new digitized and individualized manufacturing supply chains. Global customers, and not Africans, are asking for services and solutions that require high-speed connectivity for the people sitting in virtual offices around the globe. And yes, they also have money to pay for such services.

The satellite-based high-speed data link is therefore for commercial use and needs to connect the brains that provide services requiring 5G speeds. The data connection from satellites will be "tight beam," or focused "spot beam," which means it will go directly to the places where there are trained African brains instead of spreading the bandwidth across vast areas of land and people. These places are relatively few, maybe hundreds across Africa.

Each location can then establish a 5G infrastructure, and associated edge computing and VR capabilities. "Edge computing" in this context means the use of local computing devices to process, manipulate, and compress data while the virtual work is being done, before sending it through space using costly bandwidth. Such 5G and computing capabilities do not need to cover the large land areas of Africa but can be focused on a relatively small number of locations. In offices close to these locations, thousands or even tens of thousands of people now can be connected at top speeds.

During a visit to India in the early 2000s, my host and then new friend K. Rama Raju introduced me to his family's new outsourcing campuses near Hyderabad in Andhra Pradesh. It created a lasting picture in my mind, a genuine "Eureka" moment because I suddenly "got" it. The quality and even style of the facilities were on par with the most advanced in America. In that respect, it did not matter where these professionals were located.

The above concept achieves the same for Africa, but arguably on an even larger scale. Digitized global supply chains and personalized digitized production are likely to require many more people.

For this to work and fund all these investments, the economic model must be sound and scalable. We need to implement the third step also, and now Blockchain plays a major role.

Step Three: The Internet of Brains (IOB)

An initial focus on commercial use of high-speed Internet access opens the door to a wide range of Blockchain-enabled economic models and solutions.

Blockchain applications can introduce reliable ways of accounting for using data. Payment for data use would not be measured by monthly charges but be linked to the specific job being done. This leverages IoT-style accounting and billing, enforced via economic models leveraging Blockchain technologies. Similar to mobile phone minutes, the token reflects the actual data throughput used. Blockchain's additional relevance is in it being borderless and creating an immutable record, validated via whatever crypto-tech method best solves the challenge.

If an African user then interacts with a customer, the cost of using data can be integrated into the service cost. The customer pays, and the infrastructure costs should not be dramatically different from comparable cost anywhere in the Northern Hemisphere. Most of it is similar, and even targeted tight-beam Internet via LEO satellites need not be (significantly) more costly than traditional fiber optics for the masses. The business case becomes universal, and Blockchain can enforce it. Even the cost to process and validate Blockchain transactions documenting the accounting can be included into a per-job charge.

And Now for the Money

The above three steps profoundly change the business case for investors, satellite providers, or businesses establishing 5G infrastructure. No longer do they need to figure out how to get paid by low-income consumers in poor Africa. In the end, it will be the globally located manufacturers, businesses, or relatively affluent consumers that foot the bill – whether the service was delivered from Tokyo, Los Angeles, Marseilles, Kinshasa, Lagos, or Nairobi.

How much would such a network cost? Let's make a back-of-the-envelope calculation.

> If there were 1,000 innovation centers, with one dedicated satellite each, costing about $10 million each per year to provide 1 Tbps capabilities, then each one of these could provide up to 1,000 5G connections at any time, or 20,000 connections at today's broadband speeds. Assuming two shifts of work to service the digitized supply chains, between 2 million and 40 million people could then be connected at a speed of either 5G or broadband.
>
> The cost would be from about $10,000 per person and shift, at 5G speed, to $500 per person and shift, at broadband speed. Assuming 4,000 work hours a year for two shifts, the hourly cost would be $2.5 to $0.125, respectively. That seems acceptable if it results in an active involvement in the global workplace, and if the compensation level for such services is close enough to wages or salaries typically paid in the industrial world.

By establishing the commercial platform capabilities and hooking up the brains to the rest of the world, the playing field for Africans effectively has leveled. To a solution provider being part of the digitized supply chains, these brains are as capable as those somewhere else. Training, the conveying of specific skills, is something that needs to be done no matter who the people are providing the services. This can be spread across the world, and in Africa across the continent.

All this creates a new business model for our African heroes, paying for high-speed Internet connection and a linked 5G infrastructure. The rapid development model no longer needs to assume low-cost manual labor working on the assembly floor to create massive numbers of semi-finished products. Robots will take care of that. But somebody needs to look after these robots. Digitized supply chains use brains, not brawn.

It's a blue ocean that nobody has chartered before. The future of manufacturing can arrive in Africa at the same time as anywhere else.

Enabled by Blockchain technologies.

The Satellites Are Launching Today

There is one additional condition, though. The action must start now. LEO and VLEO satellites are being deployed within the next 2 to 5 years. This is a window of opportunity for Africa to immediately join in on the ground floor and actively participate in contributing to building the future of manufacturing and business relationships.

The alternative is not pretty. A failure to establish economic models as described above could keep Africa in the second or third tier. If Africans have to pay for high-speed connectivity and a scalable 5G infrastructure themselves, funded by low incomes, it will be a relatively slow process.

The worst consequence would be missing out on the opportunity to leverage this abundant resource of young, creative, and ambitious brains to transform and transcend the economic landscape while having to pay for creating the most advanced communications infrastructure on the Planet.

Using Blockchain-enabled economic mechanisms to establish the communication infrastructure outlined here has huge benefits for Africa. The most important aspect of the economic model is that global customers pay for this infrastructure to be established. Fail-safe backup Internet connectivity options may require additional MEO and GEO satellites, or even physical fiber optic cables. Although those will be part of the overall solution, most of the time they can function as a more traditional communication infrastructure for the masses of people living in African societies.

The overarching benefit for Africa, though, goes far beyond technical considerations. Africa's young are learning state-of-the-art skills and become fully integrated into the world economy and global digitized supply chains. Since their job complexity is on the medium and high-end of the scale, they will likely get compensated at a level much closer to what one would need to pay for comparable work done in today's rich world.

Money flows into Africa as investments. Skills and infrastructure are being developed. And then even more money flows into Africa to pay for high-end labor.

Money for Africa, indeed.

Chapter 18

Education. Power, Grandma Style

MY GRANDMA was right. "They can take everything from you, but not what you have in your head," she used to say. Having experienced wars, famines, poverty, and coming strongly out of each trial, she valued education a lot. You can lose everything, but what you learned stays with you and can make you successful again.

Likewise, over 75% of the respondents to our TAS2018 survey of African Blockchain enthusiasts had stressed a need for continuing Blockchain education. The responses would likely have been similar for many others emerging technologies. It makes sense. Because my late grandmother knew what she was talking about.

Even when smart African brains find a workplace with high-speed Internet connection, this alone still does not make it possible for them to work and contribute. They also need to know what they are doing, and it has to be something that their customers or employers want or need. Training and building skills are paramount conditions for succeeding in the world of tomorrow. The physical tools are meaningless unless the brains are ready and able to use them.

A Triple Challenge: $a + b + c = \infty$

The survey respondents of our May 2018 workshop in Kigali stressed no other need more than education. This gives witness to the fact that the most important and powerful raw material of the African continent is the grey matter of its 800 million young people. It also reflects a deep yearning to be prepared for the future and figure things out for oneself and Africa. This works on multiple levels.

Any basic education will help Africa advance and prepare for the future. More critically, though, the *right* education and training can drive wealth as never before. Using the Internet, learning may, in theory, become easier than ever, but picking and verifiably acquiring relevant skills is a big challenge.

In the past 30 years, the world experienced the far-reaching impact of personal computers, client-server and enterprise computing, the Internet and more specifically the World Wide Web, mobile phones, social media, and smartphones, to name just a few new technologies. We are now on the verge of a world with ubiquitous use of artificial intelligence, virtual and extended reality, Blockchain, renewable energy, 3D printing, genomics, robotics, quantum computing, and so much more. Their direct and indirect impact constantly change our lives and require new skills.

The challenge is broader and deeper, though. It does not just relate to our daily lives and jobs, but even reaches deep into science. In his book *The Half-Life of Facts*, Samuel Arbesman describes how knowledge evolves on several levels.[163] Measured by a number of criteria, like the number of scientific journals or the number of scientific papers, the output of every field of science grows exponentially. In chemistry, knowledge doubles every 35 years, in geology every 46 years, and in genetics every 32 years.

These numbers get even significantly smaller when measured by books in use, articles published, and papers cited. Based on such measurements, the half-life of correct and relevant facts is below ten years for almost all fields of science![164]

With knowledge increasing exponentially, technologies also progressing exponentially, and thus job and skill requirements changing dramatically, how can Africans even possibly get ready for high-income work of the future? Reaching the current educational standards of the rich world proves to be demanding and expensive enough. With limited resources, notably money, how can smart

Africans cope and continue to adapt and learn even more?

Educators, students, workers, governments, and entrepreneurs are all severely tested by a triple-challenge.

1. How can one effectively educate and train such huge numbers of brains?
2. How can one know what specific contents, skills and knowledge best address the demands of an unknown future?
3. And how can one ensure these skills will be used once students graduate?

After all, not only are 800 million young brains Africa's biggest asset. This also means that not developing, empowering, and productively using them would turn them into Africa's biggest waste. Nobody can afford to let that happen.

Human Brains: Africa's Greatest Asset or Biggest Waste?

Recognized as one of the best universities in the world, the Massachusetts Institute of Technology, or MIT, is one of the leaders of research about emerging technologies. Under the banner of MIT Technology Review (MIT TR),[165] one of the world's oldest technology magazines, it publishes several electronic newsletters about important new technologies. They consist of daily reports on emerging technologies ("The Download"), artificial intelligence ("The Algorithm"), and Blockchain ("Chain Letter").

One of the most fascinating aspects revealed by MIT TR is the combination of width and breathtaking speed of technology-induced change, affecting all aspects of human life. Because of that, just a few months ago MIT added another newsletter, looking at what they call the "workplace of the future." This newsletter is called "Clocking In." Providing information about technology changes is not enough, but one of the most serious implications is what it does to the future of work. Jobs will change everywhere, and most changes will be driven by technology.

The consequences for everybody on Earth are far-reaching. We all need to learn constantly, not only during the early stage of our lives. More than ever, continuous training becomes important. And this also has a direct relevance for Africa.[166]

Change always involves opportunity, too. Broad change means *lots* of opportunities. Rapid change leads to *constant* opportunities. It also levels the playing field. If everybody needs to change and adapt to something new, then nobody really can have a distinct advantage.

Africa "just" needs to find a good model and approach, at least for a substantial portion of its population. It will participate in actively shaping the future and economically benefit from that.

Education in the 21st century needs to be flexible, adaptive, with a focus on the relevant. Beyond the most basic layer of "reading, writing, and arithmetic," or underlying abilities of critical and conceptual thinking, there are severe limitations for standardized curricula in a nationwide education system. When even basic scientific knowledge and facts change and double every few years, and revolutionary technologies enter our lives even faster, we need new ways of thinking about education and the learning of skills. Otherwise, education and training may waste a lot of money, with little chances to turn skills and knowledge into profitable activities.

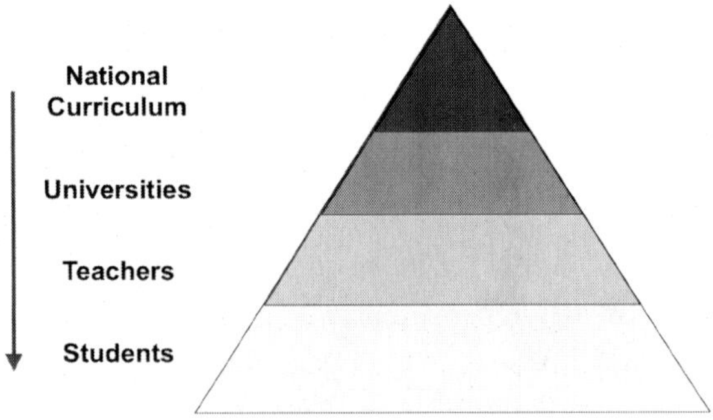

Figure 11.
Traditional Education Model

Like so many solutions leading today's world, traditional education was shaped by the addressing and overcoming of late 19th-century and early 20th-century industrial-age challenges. They are therefore dominated by centralized, hierarchical, and linear approaches. It starts with smart people putting together a curriculum, formally accrediting institutions of education, educating the educators, having the government and "society" push people into this system, and then processing students over several years until they graduate with a formal degree.

Then students can pursue follow-on degrees. At the end of this decade(s) long route they can use their degrees to look for jobs. Employers hope that the output of this lengthy process suits their needs.

50 years ago, this system worked. But as we have seen, the amount of relevant knowledge increased at an accelerating pace, and technological change kept speeding up also. The educational cycles became longer than the above described "half-life of facts," and even more so than the specific skill-requirements for new, high-value jobs.

In today's world, going into a monolithic system or even building at age ten, or later at age 18, and leaving it "educated" four, six, or eight years later – this is neither going to be "right" nor "enough." As Mr. Arbesman explained, by that time half of the knowledge will already be outdated, and likely at least one set of new technologies will already have been adopted by the most advanced 25% to 50% of the people on Earth.[167]

The traditional system is way too slow and rigid, and costly. A long-term standardized system to learn relevant skills does not work anymore. It also uses the wrong economic motivations and mechanisms. Educational institutions have little to lose even if the knowledge and skills they train are not marketable. Students (or governments, if centrally paid for via taxes) may spend a lot of time and effort, just to end up with the wrong or with insufficient skills.

So, let's turn the traditional model around.

A Continuous Sharpening of African Skills

Africa's many young brains need to be among the first to become experts in what is new.

Their education and training must be able to react quickly and consider each individual person's specific situation. Africa may have hundreds of millions of people, but no two of their brains are the same. Each belongs to an individual person, living in a specific context. Empowering means just the opposite of having centralized authorities telling everybody what to do. Rather, it means connecting each person to opportunities for acquiring knowledge and skills, guiding them, but then letting them make their own decisions on what best suits them.

Blockchains and tokens can enable all this to happen. They can define a mechanism that dynamically links teachers and trainers to knowledge and skills, knowledge and skills to perpetual students, and trained students and experts to businesses or situations in life that require the skills they learned.

Above all, the dominant unit of measurement for education has to be smaller, shorter, and more dynamic. Instead of multi-year degrees, it has to be closer to more narrowly defined skills, or even sub-components of skills. These need to become the central unit of education and training, focusing on what is in demand and marketable, and therefore well paid.

Such skill acquisition can then be reflected as tokens, and back up their value. People create tangible value each time they successfully learn new skills. This can be seen equivalent to what in the traditional Blockchain world was called "mining," although it is even more. Bitcoin miners solve otherwise meaningless mathematical riddles. On the other side, however, just creating usable skills and teaching them to people is already an asset for the community. It may still be intangible but it exists and can be used and monetized to create tangible value. That value then backs up the value of educational coins, which are far more than speculative objects.

Do you remember what my grandma used to say? Knowledge and skills cannot be taken away.

A token economy for education rewards the building and acquiring of valuable skills with tradable coins or tokens. For that achievement, each individual, and the education and training organizations, should be rewarded.[168] These tokens are documented on the Blockchain, proving exactly who acquired what skill, and how. When a person then gets hired for a job, fully employed or in the gig economy by contributing part-time or ad-hoc work, the hiring party needs to pay tokens to the educational institutions. This process can be automated and happen in the background.

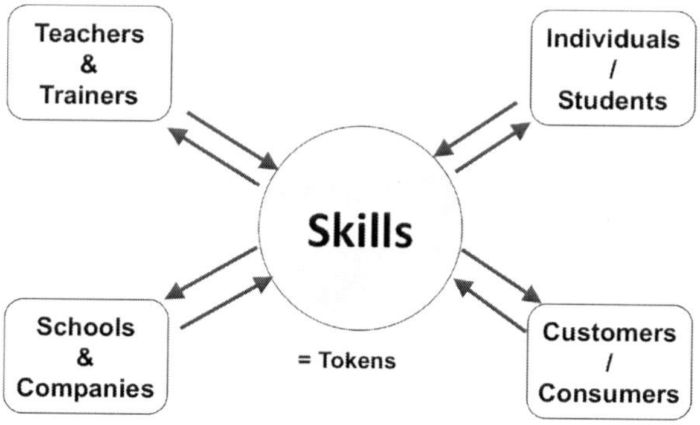

Figure 12.
Future Skills-Centered Education Platform

This links skill-based tokens to tangible value. Individuals are motivated to acquire skills since they get rewards in tokens and can prove that they have them. Businesses can purchase tokens and hire skilled personnel based on proven skills.

Now experts, students, customers, and entrepreneurs can figure out their priorities and decide on how to maximize the value of skills. The logic is straightforward. Customers and employers identify the need for new skills. The value of these skills will go up. Students now want to learn them, and teachers and entrepreneurs will develop training for it. A skills-based market can operate efficiently. One can now measure the value of one's education via the value of educational tokens, reflecting the demand for marketable skills.

In this educational token economy, market demand directs what skills to acquire. This almost gets us completely to the goal of using these skills to generate income. For this, we also need workspace, computers, and a broadband Internet connection. If then training and education are also linked to a commercial platform for executing contracting and financial payments, African brains are "in business" and can generate income.

Is the global market for such services big enough? This question can be answered in multiple ways. I simply suggest you get the newsletter from MIT Technology Review. Here is a hint: No matter how you turn it, the answer is "yes." If within the next 30 years AI alone is predicted to kill 20% of all jobs in industrialized countries,[169] replacing them with new ones, all emerging technologies together will have an even much more substantial impact.

The use of tokens aligns interests. Individuals, private educational and training institutions, governments, businesses needing skills, and investors all are active stakeholders, maximizing their own income while working for a common good. They want the token-based skills economy to succeed. At least in theory, this becomes a self-perpetuating market, a "teaching and learning machine."

From Theory to Reality

The above has too many benefits to remain a theoretical exercise. Entrepreneurs, educators, governments, businesses, innovation hubs, everybody can and probably should start exploring now. Our African heroes can work on pan-African Blockchain-based education solutions that integrate smart African brains into the global digitized world of the near future. Training up people with high-demand skills is a benefit in any way, and these skilled brains can then already actively contribute to the solution itself.

It is not a realistic option to only continue with traditional, slow, and inefficient and under-funded investment patterns. In most ways, all brains are created equal. But if we do not develop a much more dynamic and responsive effective link between future-oriented skills, educators, and their customers, then we will waste the potential of millions of our African heroes' brains.

Education and training infrastructure means more than building schools and making individuals enter to learn the same standardized curriculum. It must optimize value, and if that is done smartly, literally everybody in society benefits.

Blockchain technologies can create a situation where Africa's educated and trained minds create economic value using marketable skills. This benefits each person, African societies, and the whole world. When we start from the economic value and work our way backward, a token-based economy can achieve that in just a few relatively straightforward steps.

Chapter 19

Health Care. Making People – and Brains – Stronger

IN THE LATE 19th century, Germany introduced the concept of basic universal health insurance. Its government relented to pressure from socialists to protect workers during these heydays of the industrial revolution. Almost 150 years later, the same traditional model is still widely used. Somewhat simplified, it functions by forcing the working population to pay part of their earnings into funds, where that money accumulates as savings until it can be withdrawn at a later time to pay for medical care when people get sick.

Health Care Past and Future

Even today, many people see this old approach still worth to be replicated and even expanded on. The reality is, though, that our world and our near future bear little resemblance to the times of crude industrialization in 1883 Central Europe.[170] What we can witness today is the opening of a whole new dimension of opportunities for advanced health care. These are mainly due to progress in the fields of biotechnology, big data, mobile communication, and artificial intelligence.

African leaders have an important choice to make: Try to catch up, or leapfrog into the world of the future?

Phrased in a slightly different way, the question could be whether to search in the Europe of the 19th century for solutions to African health care challenges, or in the projected world of the mid-21st century. Or is it about who is most willing to courageously exploit emerging technologies and use a new economic and social model to battle diseases – and actively *shape* the world of the 21st century?

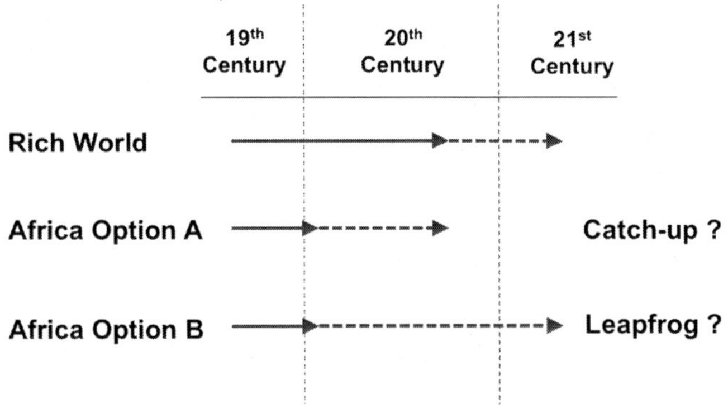

Figure 13.
Options for Building African Infrastructure

This may seem like a strange choice, considering the state of health in Africa. Some significant progress was made since the turn of the century amid a severe lack of resources. In the 15 years between 2000 and 2015, Africans' life expectancy has increased by almost 10 years on average.[171] The combined efforts of health care workers, governments, and NGOs seem to pay off.

At the same time, though, statistics and anecdotal evidence show that Africa is still significantly behind the rest of the world. Infectious diseases are the main culprits for that, although new chronic diseases are on the rise as well.[172] And so the United Nations' World Health Organization (WHO) states that "In recent decades, [Africa] has fallen behind world averages in respect of health and human development, with sobering effects on health status and trends."[173]

Un-Healthy Care

A lack of money leaves Africans in a difficult-to-escape low-health or no-health trap. Even in a utopia in which Africa would double its GDP and spend all of that money on health care, it would pay less on health care than most rich nations. Most nations on the continent are dimensions away from even achieving just that.

The challenge is on multiple levels. In many areas, there is a lack of even basic health care.[174] This alone can seem like a hopeless cause, particularly since many challenges are even bigger than in most other parts of the world. In 2015, 434,000 Africans died of tuberculosis, 403,000 people were killed by Malaria. AIDS cost 760,000, and other infectious diseases like Ebola also cost many more thousands of lives.[175] Only half the rural areas and 73% of urban areas had health clinics. Not surprisingly, then, the citizenry gets impatient. Negative evaluations by Africans have increased between 2005 and 2015.[176]

Although NGOs, among them large and well-funded foundations like the Bill & Melinda Gates Foundation, spend lots of money and are achieving meaningful results, the overall progress from these investments is slow. The uphill battle everybody faces can be overwhelming. In 2010, the average total health expenditure per capita in African countries was $135. This compares to $3,150 in average high-income countries,[177] almost exactly *double the total GDP per capita* of sub-Saharan countries![178]

In an ideal world, Africans should have access to the benefits of modern technology and to personalized health care also. To truly empower, connect, and unleash 800 million young brains, they cannot be incapacitated and struck by disease. It is that simple.

In reality, though, technological shortcomings just seem to add to the challenges. The lack of a state-of-the-art computing and communication infrastructure makes attempts to catch up even more difficult. They manifest themselves in less Internet access, low data transmission rates, and immature hardware infrastructure. The medical community then spins its wheels. Physically storing data on other continents may be costly although technically feasible using cloud services. But then data transmission becomes an insurmountable bottleneck.

Even if data similar to many rich nations is collected, it is fractured and gathered in a disjointed manner, often not using scalable electronic means. Much of this data is unreliable. Weak identification solutions prevent its proper collection, analysis, and use. Hardly any African genomics data has been collected.

In light of all of this, how could crypto-technologies even just theoretically help to make a significant dent? It seems unreasonable, hyperbolic, maybe even delusional to talk about leapfrogging when one is so far behind.

The above frustrations and the great aspirations, though, let one also wonder why there would be any objection to trying a new model. Maybe spending astronomical amounts of money on merely attempting to close a gap to where other nations were in the past (and where they eventually hit a wall anyway) is not the answer. A lot of the money could be better spent on future-oriented solutions.

And just maybe, these solutions may cost less, and even fund themselves. A look at first principles is necessary. Following the strategic planning method called "backcasting," we can visualize an ideal desired future and work our way back to see what it would take to get there.[179]

To do just that, the first step begins with looking at what potentially helpful new and emerging technologies and capabilities are available.

Emerging Health Care

Approaches using token-based Blockchains, combined with other emerging technologies, can drive solutions to Africa's health care quandary. They may even be indispensable. Blockchains are by definition decentralized and trustless, meaning that one need not put all faith into centralized organizations or businesses. By creating immutable records, it makes it easier to find one's health records and biological data, while reducing the likelihood of misuse. Storing such data in distributed form, enforced by Blockchains, is argued to be significantly cheaper than storing them in silos or in the cloud. Encryption can protect confidentiality and prevent misuse.

From the perspective of the person seeking health care, even the cumulative effect of all these benefits may only be an incremental improvement. It pales compared to the much more far-reaching and genuinely transformative impact that introducing new crypto-economic models can have. By developing superior motivation and reward mechanisms, they can introduce a threefold attack on current constraints.

- **Health Care can be pre-emptive and preventative.** The traditional after-the-fact approach centers on treating people when they are sick. This catches people at their weakest while also undermining the support networks that must take care of them. The associated economic and personal costs are huge, and the solution is inefficient.
- **Health Care can focus on each individual person.** Most of today's care follows generalized approaches that have obvious limitations since they can only pay insufficient attention to the fact that each human is different.
- **Health Care can scale out quickly.** Today's complicated bureaucracies and inefficient data collection still keep the focus on manual-labor intensive work. Investments and health results currently have more of a linear connection rather than a much-preferred exponential one.

Each limitation can be addressed by introducing new token-based economic motivations and solutions. This requires a profound reassessment and redesign of traditional thinking about health care. It also may only have become realistic with the invention and spread of genomics, mainly lower-cost high-throughput gene sequencing, and gene editing tools like CRISPR Cas9.

Health *Assurance* Over Health Insurance

Preventative personalized health care tapping into global AI-based benefits is the ultimate decentralized approach. Its basic idea is to prevent people from getting sick by addressing lifestyle choices and identifying genetic dispositions before they manifest themselves in diseases. Economically, and comparing the same net results, the "health assurance" costs over a human life should be lower than money that otherwise would need to be spent on "health insurance."

Such an approach is tempting but also seems realistic, or at least worth to seriously explore, since even the rich world faces huge challenges with its contemporary health care models. Per capita, the world's most indebted country is Japan, to a significant degree because of the rising health care costs associated with a rapidly aging population. Also per capita, the United States spends twice as much money on health care than anybody else and the overall amounts are rivaling the size of the whole budget of its federal government.[180] Even European health insurance systems are under great pressure and frequently rationalize health care.

It can be a nearly insurmountable challenge trying to reform a sector so entrenched and politicized, and so emotionally charged. For Africa, however, this is an opportunity to pursue deliberate strategies that transcend and leapfrog, instead of trying to emulate what inevitably seems to lead to a dead-end anyway.

But is it even possible?

There are in principle three ways that Blockchain technology can provide value to health care in Africa. They can facilitate an improved storing and sharing of traditional health care related data, enable preventative health care using genomics, and transform the traditional health insurance system via innovative token concepts that fund the whole effort.

All three require a shift in our thinking, breaking away from ingrained patterns.

Shift #1. Blockchain-Enabled Collection of Individually Owned Health Data

The beginning of the process is to acknowledge that each human's medical data should be owned by them and not by medical or health care institutions. Far from criticizing the roots of the current approach, individual ownership of data is crucial to establish economic motivations that fund future efforts. Today this health-related data does not just grow exponentially but its value does also. Individuals should be able to benefit from anybody coming up with treatments or drugs based on analyzing their data. If so, people are directly motivated to collect and contribute more of it, creating a positive feedback loop. It would be their personal choice, though.

Associating data with a token establishes the foundation for a market in which such information can be traded. It also provides the link for benefits from research and development success to flow back to the individual that provided the "raw material," the data, in the first place.

This can only be done with a powerful identity solution. Once an individual is identified, all medical data associated with this person can be encrypted and linked to the person, documented on a Blockchain. The data being covered can range from specialized medical X-Rays or blood tests to general phenotype information like gender, family history, or environmental conditions, and reaches into lifestyle tracking of activities using mobile devices.

As of the late 2010s, data from wearables, sensors, and social media increasingly keep entering the world of health care. Many initiatives try to figure out how to integrate it into the overall patient picture to provide a more comprehensive understanding of triggers and influencers of health and diseases. Blockchain cannot just make this easier, but it may be a key tool that has to be used to protect privacy while providing flexible access during that process, particularly when AI is analyzing health data.

Traditionally, data in health care consists of electronic health information that usually is collected in a doctor's office or hospital, and in related administrative records like those required for prescriptions and insurances.

Patient visits can create a large volume of data, like Electrocardiogram (EKG) images, X-Rays, or claim and compensation forms. This data is stored locally and in silos not immediately or easily accessible by other interested parties, whether they are patients, researchers, doctors, or people involved in claims processing.

Health data could in theory physically remain stored in the cloud, with local health care providers, or even inside a person's body.[181] But Blockchain-enabled decentralized data storage offers powerful benefits. It can make the process more efficient, better address privacy and data security, and still enable AI algorithms to analyze the data in the search for correlations and causations leading to new treatments or even cures.

Costs for data storage may also be lower with Blockchain projects. SIA and Filecoin claim they can significantly lower storage costs, sometimes by a factor of up to 100.[182] For the users, it would be an effect similar to using cloud storage. If African health care providers are looking for effective storage solutions, the likes of SIA and Filecoin, or other solutions based on Protocol Labs' InterPlanetary File System may be the way to go.

The arguably most significant data component can be a new one: information about one's DNA. This introduces a new dimension of opportunities.

Shift #2. Leverage Genomics

Although the biotech revolution brought about by genomics is still in its infancy, it promises to rapidly upend health care as we know it. The basic logic is the ability to identify for an individual person the likely causes for diseases that have not manifested themselves yet, and to use gene-editing based treatments to prevent these diseases from breaking out in the first place.

Once the first human genome was sequenced around the year 2000 with an effort of multiple years and at a cost of billions of dollars, both costs and speed dropped exponentially. Nowadays a human genome can be sequenced within mere hours at costs below $1,000.[183] In just a few years this will likely be possible in minutes, costing less than $100. This is the future that Africa needs to prepare itself for.

The term "genome" describes the genetic building blocks that determine the physical characteristics of each human. It consists of a sequence of several billion letters (C, T, A, and G) that can easily be digitized. Diseases or likelihoods for diseases can then be identified by little "bugs" in the genomic code, like switched, omitted, or added letters. Critical conclusions can be drawn by having AI algorithms compare large numbers of genome data between sick and healthy humans, and correlate them with the above-described more traditional phenotype and newer lifestyle data. Most importantly this can point at causes for diseases and suggest treatment options.

One of the main challenges with genomics is the huge data volume and the still relatively high costs associated with sequencing. Although its quality gets better, sequencing has not been perfected enough yet. Treating or preventing diseases will require continuous genetic analyses, further increasing the data volume. All this costs significant amounts of money, mostly for computing, data processing, and data storage.[184]

Like with traditional health data, sequenced genome data should also be stored under the control of individuals, encrypted and enforced by Blockchain mechanisms. With the much more significant potential to lower overall health care costs, genome data becomes even more valuable.

The generation, storing and sharing of genome data involves a number of privacy, security, and ethical questions. Their discussion goes beyond the scope of this book. As a general statement, though, I see gene sequencing becoming as ubiquitous as face recognition. The big issue comes when a centrally controlled database can be misused, and this enables massive-scale attacks. On a personal level, though, we have to assume that complete privacy cannot exist for any person anymore. Similar to a fingerprint, genome data can soon be gathered at a low cost from a hair or a glass that a person touched. However, since the identity of a human goes far beyond what even a fully sequenced genome can reveal, privacy and security need to consider a much more comprehensive picture.

Precision medicine for all seems like a First World dream, but a smart use of technology can make it a realistic aspiration for Africans also. What it will entail is the setting up of African sequencing labs that can flexibly leverage edge-computing capabilities like those I have outlined in my analysis of an African communication infrastructure. A great option would be to put such processing capabilities in innovation hubs located across major cities in Africa, close to where 5G-level data transfer and edge computing capacities are established.

The main challenge is cost, and this is the number one issue for health care anyway, in Africa and beyond. To effectively address costs requires the third and most important step of the plan: a new economic model to pay for African health care.

Shift #3. Use Tokens to Replace Health Insurance with Health Assurance

At today's rate, it would cost over $1 trillion to sequence all African's genome, about six times as much as all 54 African nations together spend on health care.[185] This data then also needs to be processed and mapped, probably at least doubling the costs. Then there is additional contextual "phenotype" information, traditional health care data plus such from wearables.

The real dimension of the problem becomes apparent. Africans just cannot spend 10 to 20 times of today's annual budgets on the health care of the future while much of the population has no access to modern health care yet, and the majority of the others have it only at a low level. Or can they?

To just have any chance, a new economic approach is needed. It has to economically motivate private investment to pay for collecting and storing health-related data. There needs to be a clear path to reap dividends from these investments within a reasonable time frame.

Once that is in place, spending money to accumulate health-related data makes sense as a form of investment.

As more data is being collected from ever more people, it gets easier to use "big data" AI algorithms that find correlations and causations. This increases the value of the data over time while turning it into tangible and marketable value, in the form of treatment options. As ever more data is collected the cost of that effort goes down. Economies of scale kick in. The costs of gene sequencing and mapping, data storage, and bioinformatics and AI-based data analyses drop.

Blockchain is the secret sauce that makes all of this happen. A good example is the Blockchain-enabled initiative Healid that I co-founded. It aims at the rapid accumulation of genome data. To achieve this, we are turning the economic model of traditional health insurance on its head. Our goal is to enable a market for future health via tokens. A token is rewarded for personal health care related data, like files containing gene sequences. To access this data for research or treatment, a price in tokens has to be paid.

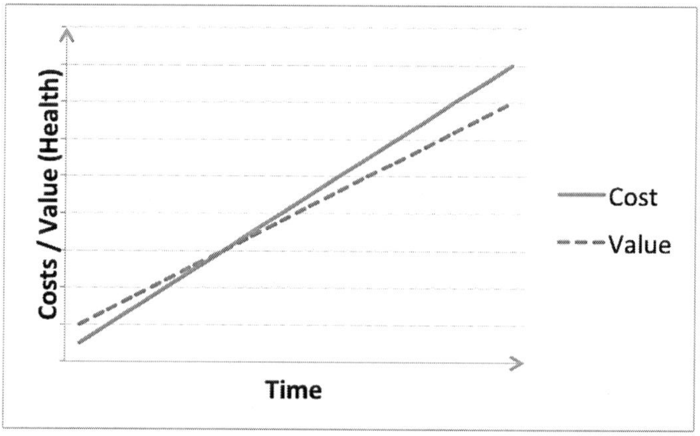

Figure 14.
Health Insurance: Traditional paying for health via insurance.

Such a Blockchain-based approach requires three components that mutually reinforce each other: (A) the collection of genomes and phenotype data, (B) the sequencing, mapping, and storing of genome data, and (C) money to cover the costs for collecting, sequencing, mapping, and storing. The basic logic is to give a token reward to everybody who contributes any of these three components. To access the stored data for research, development, or treatment, payment via a token is required. AI-driven analyses of the data find new causations that lay the foundation for treatments of diseases. This releases the prospective value of the data via personalized preventative health care for all humans. Not just Africans, but patients all over the world benefit. They also can use the token to later pay for such futuristic health care.

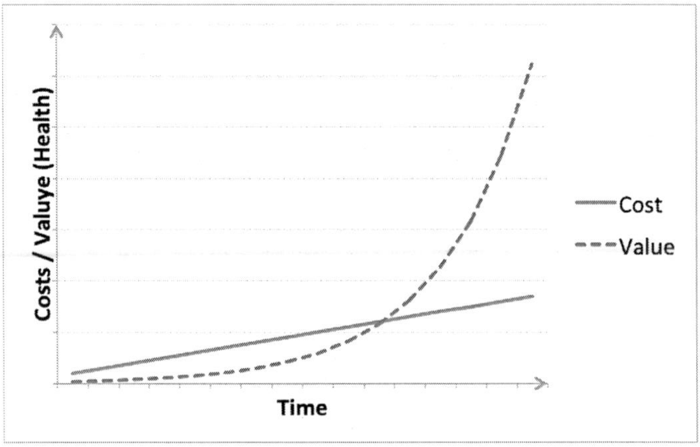

Figure 15.
Health Assurance: Future paying for health via investments.

The approach uses market forces only. Africa has something immensely valuable for the bioinformatics industry that uses such data to develop treatments for diseases. It has over one billion sources of human DNA. Collecting and analyzing this data in Africa or anywhere else helps everybody on Earth. It is why our slogan at Healid is "Heal One Another."

Obviously, sequencing the genome of all Africans is not even possible yet anyway due to a lack of sequencing capacity. As of mid-2018, the total global gene sequencing capacity is below 10 million, maybe even below 1 million.[186] Proper motivations can create an increasingly large volume of individually owned but universally accessible health-related data. The value of the database increases with more data being stored, driving the value of the tokens, and reducing the costs for data collection. It becomes everybody's self-interest to make data accessible, ideally in encrypted and not directly attributable form.[187] A virtuous cycle has been created.

Taking care of basic health care needs for everybody may be the most immediate and from that perspective the biggest challenge for Africans. The most far-reaching opportunity, though, is in more profoundly transforming the way we think about health care.

When the currently dominant health-insurance system was invented and established, the world looked different. In the 1880s, the combustion engine was invented but there were no automobiles yet. Edison invented the electric light bulb, but electricity was not yet commonly in use. Telephones were just emerging as a tool of mass communication, data was still mostly stored on paper, and Louis Pasteur and Robert Koch led the emergence of immunology and microbiology. This was by far preceding mainframe computers, personal computers, the Internet, mobile phones, satellites, 3D printing, solar collectors, and so much more.

All these technologies have evolved substantially and are being replaced or profoundly transformed today. Similarly, it is about time to move beyond the 150-year-old approach of using health insurances as the main basis to finance research and development and to pay for health care.

Africa can do it by flipping the logic from saving to investment, and from insurance to assurance.

Blockchain-Enabled African Health Care

Right now is an excellent time for Africans to take action and move into the future of health care.

Imagine people not getting sick, because they can be treated in advance. Imagine a source of asset with rapidly increasing value shared for analysis with the rest of the world. Imagine the provision of advanced health care data analysis services out of African innovation hubs.

Once the health care data of millions of Africans is encrypted and stored under each person's individual control, but accessible via Blockchain-powered security and new identity models, and their genomes are sequenced, and AI can run over this data – then money can be spent on powerful preventative care rather than "wasting" it after the fact once people get sick.

Our initiative Healid has the motto "Heal One Another." This is what Africans can do if they embrace new health care technologies and economics, enabled by Blockchain.

Chapter 20

Energy.
Power to and for the People

THERE IS NO REAL POWER without power. Our world needs electricity and even more generally energy to keep the lights on, power computers, drive machinery and move cars, trucks, and buses. It, therefore, is high on Africa's priority list despite the slow speed of progress in what seems like an eternal uphill battle.

Again, this is about infrastructure, and again it seems like one more of these vicious cycles. Without enough, affordable, and reliable energy, productivity is low and economic output is small. This then leaves little room to invest in costly power supply, storage, and distribution infrastructure. And so, the cycle continues in rapidly growing and ever more power-hungry societies that are cash-strapped, preventing Africa's leaders and brains to play on an equal footing with the rest of the world.

How can African leaders rapidly empower 800 million young brains, so they can contribute to building a bright future for humanity in the mid-21st century?

But this asks the wrong question. Wanting to catch up is the issue. And the leaders need not do it anyway; they merely need to enable it. A fresh look at creative options and new approaches may enable community leaders, entrepreneurs, and political and business visionaries to break the old cycle and turn vicious into victorious.

Again, Blockchain-enabled economic models can play a helpful and empowering role, in more than one sense.

Dark Africa: Power for the People

At night, Africa is dark (see picture below). The home of the Black Panther looks like a genuinely black continent. Whereas East Asia, Europe, North America, and even India are light, a satellite image of Africa makes it hard to believe that over one billion people are living there.

Figure 16.
A Map of Africa and the World at Night

In general, despite the continent's wealth of natural resources, Africa only produces insufficient amounts of energy. Unlike in the rich world, electricity is not readily available to its businesses and to its people. Most African countries do not have a large-scale and reliable power grid. Resolving this issue is difficult. What industrial countries built over a century proves a lot to bite off for cash-starved economies in a matter of several years or a few decades. It took most of the industrialized world 150 years to reach their current levels. How could one possibly catch up in a much shorter period of time?

Investments in traditional forms of energy production and

distribution are very expensive. Dams, fossil or nuclear plants, power stations, and transmission lines all cost lots of money, require substantial engineering efforts, and take time. It seems a losing proposition.

The traditional solution is for African governments to extract taxes, borrow funds, or use development aid, and then invest this money into building the power infrastructure and electricity grid that industrial countries have. But it simply seems ever more likely that emulating hierarchical industrial-style approaches and solutions is not the answer.

The home of the Black Panther should be as bright as the rest of the world. And maybe this fictional hero can give pointers. Above all, he lives in a world of the future and in a world of technology.

Let's dive into those.

Solutions: Power for the Black Panther

Elon Musk is a technology visionary from Africa, a self-made multi-billionaire. He also understands a lot about energy. So, it is notable that while politicians in the U.S. had been advocating political solutions to North American energy independence, Mr. Musk did something similar, but with a crucial twist. He promoted energy independence for each *household*. Generate electricity with solar panels on your property, store it in a wall battery, and power your appliances and your car with it. You will not need power from a utility or gas station any longer.

His vision is transformative because his answer focuses on empowering individuals and small groups, maximizing the strength of decentralization. Empowering individuals can be a highly effective way to solve society's problems.

Hold on, did I just say "Blockchain?" Decentralization and leveraging the power of peer-to-peer networks without centralized control that is what Blockchain and crypto-technologies seem all about. Although the term often is misunderstood and misused,[188] "decentralized" is probably what most Blockchain enthusiasts would say if they only could use one word to describe what Blockchain is about.

For transformative success, though, it is more appropriate to look at the situation as a clean slate. Since something new has to be built anyway, it is worthwhile looking at what that new would look like in an ideal world. This is a first principles approach. It focuses on the tools and capabilities at hand, instead of existing and potentially outdated processes and sub-optimal solutions.

New technologies like batteries, renewable sources of energy, and smart grids point at superior futuristic options. Blockchain can pull all these aspects together via a creative token mechanism. New economic models and organizational principles can attract funding, and so contribute to the more efficient and less costly building of the energy infrastructure that African societies need.

The lack of a sufficient energy grid seems similar to the lack of a widespread telephone landline infrastructure. The African response may be similar. It may be an obvious case for Africa leapfrogging in the energy sector also, rather than trying to replicate rich countries' infrastructure.

This can be achieved in three steps.

A. Identify Futuristic Solutions That Fit Small Communities.

Most rich countries' traditional solutions use petroleum, natural gas, and coal to produce energy, complemented by nuclear energy and hydroelectricity. Being massive industrial and logistical endeavors, they require large power plants, refineries, pipelines, intricate shipment and distribution channels for fossil fuels, as well as extensive transmission lines forming core components of complex electrical grids.

Several trends of the energy sector are profoundly challenging this model:

1. **Sources of renewable energy** are rapidly getting more efficient and less costly, particularly solar and wind.
2. **New and revolutionary battery technologies** provide improved abilities to store energy.
3. **Micro-grids and nano-grids** enable communities to manage the production and distribution of energy on a local rather than a regional or national level.
4. **Battery-operated electrical devices** are spreading and permeating societies, including most means of transportation (cars, trucks, railroads, and ships).

The future looks different than the current state of the industrialized world. A comparably massive energy infrastructure is not needed anymore. There is no need to replicate centralized and fossil-fuel-based solutions rooted in the industrial revolution.

> Again, here is a back-of-the-envelope calculation. According to the International Finance Corporation (IFC), a subsidiary of the World Bank, in 2014 an average household in Ghana used 744 kWh, which is 1/16th of the average household in the United States (12,305 kWh).[189] One large solar panel can generate enough energy to power a Ghanaian household. Ten solar panels and one Tesla Powerwall battery can be purchased for less than $150,000, enough for ten households. Amortized over 10 years, this would result in about $1,500 per household and year. That is $125 per month or $4 per day. The per-household numbers for wind turbines would be comparable, although one such turbine could power 8,000 Ghanaian households.

Technologically, Africa can leapfrog. To actually do so, however, it also needs the proper economic principles and mechanisms to support such decentralized approaches.

B. Link Blockchain Tokens to Value: Electricity.

The power grids of the industrial world are for energy what telephone landlines are for communications. They are expensive to set up and maintain, complicated and sometimes even dangerous, and they provide a 19th and 20th century-style solution that works but is not up to what the future brings. Africans can jump right into that future by embracing decentralized energy production and distribution.

Blockchain-enabled micro-grids enable small communities to produce, store, distribute, and use electricity locally.

> "Coupling cognitive solutions with intelligent grids will enable consumers and energy distributors to work in partnership, predicting peak consumption and locating surplus energy without demand, thus avoiding blackouts and other problems posed by the traditional grid networks."[190] (Irene Sun, KiWi New Energy)

A look at energy storage can expand on this point. Batteries are effective in reducing electricity demands. Today, huge amounts of power are wasted since supply and demand do not match. Electricity is created at different times and different locations than when and where it is used. To account for peaks and losses, electrical power must be created in excess of demand, or it will only satisfy low demand and become insufficient during times of peak demand.

Lowering energy requirements is far cheaper than increasing electricity production. Smart grids use AI to match demand and supply *over distance* and across more consumers, while batteries add even more scalability by balancing peaks and valleys *over time*. They can store surplus electricity when there is low demand, and release it at peak times. A token can be an effective accounting tool, representing units of energy created, stored, and used.

Nikolay Martyniuk, co-founder of Blockchain startup WePower describes the opportunity as follows.[191]

> "As the world moves away from using coal, gas, and petrol into more renewable forms of energy, Blockchain becomes the best way to connect renewable technologies, devices, and grids, to people and communities. Blockchain, which functions as a public ledger or record, can take inputs like the amount of energy produced from smart devices like solar panels, record them, assign a price and then send it out to smart homes on the grid while recording incoming payments for energy purchased."[192]

Blockchains can improve efficiencies, lower costs, and provide transparency. Smart contracts can optimize prices and enforce the use of mathematical rules for trading electricity. Middlemen cannot control and potentially distort the process. This can be achieved on various levels, so either between utilities inside a country on a large scale or between individual consumers on a micro-scale. It effectively establishes decentralized smart grids with superior performance.

IoT-like applications can provide automatic device-driven feedback and lead to new billing mechanisms. Using electricity to power a device can then work and be charged for, like using data on a mobile phone.

This opens the door to another powerful concept: letting the beneficiary of a device or machine pay for the human working with it. Is electricity used for a household item? A computer or phone? Or for a car? Neighborhood batteries could turn into the equivalent of neighborhood gas stations. For agriculture, the electricity costs to operate machinery and to transport produce could be handed down the supply chain, even to consumers.

The opportunities are sheer limitless. Smart, creative, driven minds can take advantage of them. As we all know, those can be found in every building and at any street corner of Africa.

C. Scale Out Via Token-Based Investment Models

So far, I have outlined self-contained business models based on future-oriented energy tech, preventing waste because they do not blindly copy outdated solutions.

The third challenge is to develop the economic motivations that promote and entice investments on the massive scale that Africa needs. Our overarching purpose is to provide as much energy to Africans as possible, thus enabling and driving economic activity and advancement. As energy production increases and economies get wealthier, the value of energy output increases, and so does the value of tokens connected to it.

Smartly scaled out, installing and supporting micro-grids, or nano-grids do not need to be substantially more expensive in Africa than in North America. The relative incremental benefits for Africans are substantially higher, though. Going from 10 to 20 is a bigger relative step (100%) than going from 150 to 160 (7%). Or, to stick to the previous example, adding 1,000 kWh to the electricity consumption in Ghana is an increase of 134%, but for the U.S. just 8%.

Tokens can align interests of investors that fund energy generation and batteries, with this equipment's operators, and the local community. Investors, operators, and users become co-owners. By linking the token to value created by generated electricity, this can even establish a consistent revenue stream for everybody.

The investment can then be treated like a typical startup. Early-stage investments get preferential treatment, with higher token rewards. This does not work well on a small scale, because it does not make sense to invest in a token that only facilitates the operation of one small micro-grid of 10 or 20 households. Such an investment would be specific and very targeted. Small numbers make it risky for investors, and more erratic when assessing costs on a per unit basis.

Scale can address such concerns of risks and costs and establish a more substantive value backing up a token that is used widely. Risks are balanced and reduced if many investments share the same token infrastructure, ideally even in multiple nations. Economies of scale can then also drive down costs. Investors can buy tokens representing generic kWh, rather than specific solar panels or micro-grids. Early stage investors can be rewarded disproportionally. A smart algorithm can make it likely that the token value will increase as Africa develops. Indirectly, the positive effects of providing more energy to more people will result in economic development, more income, and higher token value benefiting everybody who is part of the Blockchain-enabled energy community.

Today, as small and poor markets with inadequate financial and commercial infrastructure, African nations provide no good rationale for investment. Token-based solutions like the example just outlined can change this situation dramatically. They transform the funding mechanism. They also cut out intermediaries, lower prices, align interests, provide transparency, reduce risks, and open new sources of future income. Payment in tokens may become even more acceptable if investors (grid builders or pure financiers) can speculate on the compounded increase of energy demand and future token values, and particularly if they can influence such an increase in value.

Investments in a future-oriented energy infrastructure can become self-sustaining.

Real Empowerment

Leapfrogging again is more likely the answer than replicating the century-long effort of industrial countries in building their energy grid infrastructure. This will not happen by itself, though. Communities and entrepreneurs must act. Support from governments can help, although mostly by facilitating, not by driving what have to be decentralized and privately funded efforts. However, governments may become active investors and, as token holders, thus stakeholders in such ventures.

If nothing is done, a poor and fragmented African energy market will continue to limit what African economies can achieve. It will leave Africa's entrepreneurs with at least one hand tied when competing in the global economy.

On the other side, new crypto-economic motivations and funding mechanisms can inject billions of US Dollars into African economies, shifting responsibilities from governments to individuals, and empowering societies and Africans in that process.

Chapter 21

Transportation. Accelerating Africa

5,000 MILES. 8000 KM. Africa is no mere Europe. And no mere USA or China either. These numbers are the approximate North-South and East-West dimensions of Earth's second largest continent. Traveling such distances takes time and money. Moving goods and money across Africa is slow and costly.

But people and goods need to move, and for that, they need transportation infrastructure. With limited money, however, how can Africa fund investments that took industrialized countries decades to build? Money is the big constraint. African leaders struggle with this challenge, arguably even more so than most other societies since they have so many conflicting investment needs and so little money. And African entrepreneurs and ordinary people need to live with these shortcomings. It is just another one of the competitive disadvantages of doing business in Africa.

Transportation is about moving people and goods on roads, but also railroads, ports, airports, and waterways. Traditionally these are projects conducted by governments, often on a large scale. This part of the infrastructure affects almost every single person. For many commercial activities it is indispensable, and so it is for people's daily lives.

Stop. And Go!

Considering the enormity of the challenge, priorities are important. For example, Africa does not need to rapidly build infrastructure that connects the whole continent. Any focus on a pan-African road system would be way too expensive at this point, even though it could in theory help to better integrate its economies. It is not an immediate priority, however. For the time being, most revenues from African's business activities originate outside the continent. Also, chances are that future customers and revenues will come from digitized services and solutions.

Most Africans live on the coasts and major waterways or lakes, anyway. People do not use roads to go from Cairo to Dakar. If Egyptians want to travel to India, they take a plane. They will not drive. It is the same if they want to travel to Dakar or Lagos. Likewise, the transport of goods over long distances will likely be taken care of by railroads or ships.

These are not dramatically new insights. Where they become important is to identify the correct priorities. Infrastructure is not about building monuments but solving people's problems and helping them taking advantage of opportunities. It then makes much more sense to look at the lowest hanging fruits. This means to first focus on cities, countries, and then regions.

Even then, it may seem that governments or foreign aid are the only realistic sources for building the core pieces of roads and rail, airports, and seaports. The magnitude of the effort and its associated cost can hardly be paid for by cash-strapped societies.

This changes if we consider dynamic economies. Business cases for investments significantly change if rapid economic growth of 10% or even more per year is possible. 10% growth over 10 years ends with an economy 2.5 times its previous size. 15% growth quadruples the size of an economy in just 10 years.

If at least pockets of Africa seem able to deliver that, then one can forecast the demand for transportation of goods and the movement of people, and the ability of the local economy to pay. It then comes down to developing a proper revenue model that can bridge time and can generate a good ROI via a long-term revenue stream.

Introducing Tokens

What then are our heroes of Africa trying to achieve? They want people and goods to move efficiently. Ideally, a need implies a business case for private investment, without requiring involvement of a central authority. Preferences and needs should be indicated directly by people that want to move, or by the businesses that transport goods, and by the communities that are directly or indirectly affected.

Preferences can be clearly expressed with money. The more people will pay over the lifecycle of an asset, the better the ROI and the business case for investment.

Two things are important for this. First, we need a good and consistent measurement for the value created via investments (expressed in tokens), and second, we have to make sure these investments are governed by a community of stakeholders that goes beyond the investors, the people with money. Beyond these two points, we can also scale out further by introducing a pan-African dimension, and by adding further economic motivations.

Blockchain-based token concepts may help exactly with that. They can record relevant transactions, document them on a shared and unchangeable database, and tie them to a token market.

As an example, for moving people, our heroes of Africa can establish a token based on person/km, or "pkm," and container/km, or "ckm." If a 10 km road is used by 1,000 people that travel on average 2 km per day, this translates to 2,000 pkm, and the overall "people load" (pLoad) is 200 per km. If the number of people increases to 8,000, traveling on average 3 km, then this means 24,000 pkm and a pLoad of 2,400 per km. The equivalent measurement for transporting goods can be a unit expressing containers per kilometer, establishing a "container load" value, or cLoad.

The same logic can be used for any infrastructure component that enables the transporting of people and goods, like ports or canals. It establishes the basis to compare different investment options for transportation, comparing rail to road and water and air. Now Africans have critical input to build, basing their business cases on ever more people using these roads and increasing the load. They have converted the value of moving people and goods into a token expressing that value in a consistent manner. This makes different infrastructure investments directly comparable, even across regions.

The main new concept here is the definition of a token, tied to the economic value of transported people or goods. It is important because it can be used to forecast future revenue streams. From here on creative solutions can go in many directions. African entrepreneurs can build roads as major thoroughfares for the sole use by public transportation and electric vehicles, and for the transporting of goods. Movement will be measured via RFID of NFC,[193] or similar mobile tools, likely based on 5G.

To use the road or train would then require paying a price in tokens. The price of transporting a good can then be handed on to the recipient, or to customers. If time is not important, you may use slow roads. But if you buy tokens, you can pay with them to use newer and faster infrastructure. In that respect, it is similar to the concept of traditional toll-booths. Prices can be included into bus tickets (or train tickets) and collected via RFID. AI can help to identify and minimize fraud. The exact same approach can be used for transporting goods.

Since token ownership is being recorded, ideally on a Blockchain, their use is documented also. As revenue comes in, the inherent Blockchain logic can compensate investors, community members, and all stakeholders, in an automated way similar to how IoT transactions can automatically be paid for.

Strategies to Move Faster

Ideal transportation investment projects are "additive." For example, they can separate general use from specific use. The new transportation systems can then be geared toward public transportation (e.g. using a bus or train), transportation of important goods, or cars owned by the affluent. Such a project should mainly be driven by usefulness identified by revenues, and not be a social measure. An example for the difference is a road through a city that everybody uses for "free," versus a tram that can be used by paying passengers only – or a road that, similar to a tram, only can be used by vehicles that pay either for transporting passengers or goods.

A multi-step strategy can even further involve Blockchains and new technologies on several levels. First, the highest number of people moves inside cities, or close to them. Since these cities also are the places where boost centers operate (hopefully soon), a token mechanism can fund activities, therefore pushing compensation for digitized services directly into communities.

Part of the compensation can be in convertible tokens, to be exchanged for investment stakes in local infrastructure projects. This would be instead of paying taxes. Part of one's income from external services may be given as transportation infrastructure tokens and fund such investments. These tokens enable participating in the upside by distributing revenues. Therefore, they are not lost. They can also be used as compensation to the community for the building of infrastructure, while still leaving the payer (e.g. the worker) with the chance for an upside. Effectively, this then functions like "built-in" or "incentivized" savings used for investments. It can be the mechanism that translates the income from high-tech services in the global digitized supply chains into investments that benefit local communities, and therefore the general public.

There will be obvious limitations of this. Some of the infrastructure needs to be public, meaning some roads will be used by everybody, with motos, cars, bicycles, or even to walk. Token mechanisms can still help. They can fund building such publicly shared "free" roads.

As a second layer, national or regional infrastructure can be financed and built. Such investments can appeal to general investors, including foreigners. By using common tokens based on pLoads or cLoads, investment decisions can be standardized, even across cities, nations, or even regions.

Run: Expanding from There

The above laid the foundation for privately directed investments into local and regional infrastructure. If the same token were used across Africa, then investments could be balanced based on wherever they result in the highest ROI. This spreads risk. It also creates competitive pressure to encourage people and businesses to use investments (e.g., roads), and to identify new profitable opportunities.

One can now differentiate further. For example, electric cars may be treated preferentially. Local neighborhoods may receive an allowance or distribution from the road revenues. People maintaining the road, or providing services, may be partially compensated with tokens. Transport tokens may also be linked to gas stations or power stations for electric vehicles.

Substantially different approaches may also work. E.g., using electrical vehicles may enable the free use of roads. Essentially, the cost for road usage would then be paid for when charging the battery.

Beyond the "linear assets" of roads and railroads, many investments also include stationary fixed assets like railroad stations, airports, and ports. Similar concepts can be applied there, measuring use by person and container (or equivalent) and using a token mechanism to spread the revenues.

All the above are suggestions to indicate how a new thinking and use of new technologies can create mechanisms encouraging private investments into transportation infrastructure. These are not final answers. However, they show how a creative use of new technologies can address the big-cost aspects of transportation also.

Such solutions were impossible before Blockchains, tokens, mobile LTE and 5G communications, or advanced computing. Now they are. In summary, solutions are based on these two concepts.

1. **Link products to value by defining a unit transcending borders and time.**
   The value must be based on economic activity, not purely on investment value. This enables a direct comparison between

options and motivates investments into areas of high benefit, e.g. roads in one country vs. trains in another one. Stakeholders should get higher rewards the more people use infrastructure. This can be achieved via built-in token rewards. The goal is to invest into roads and railroads or other infrastructure in a way that over time a constant stream of income is generated. It is to both anticipate and drive economic activity. The more people and goods move, the higher will be the rewards for anybody who is invested.

2. **Spread ownership via baked-in mechanisms.** Identify and reward a diverse set of stakeholders, including local communities, thus creating a waterfall effect (not a mere trickling-down). Aim at establishing a balance between financial investors and other stakeholders. Several general rules need to be balanced. (a) The simpler the logic, the better, since everybody is clear about motivations. (b) The wider the net of stakeholders, the higher is the probability that a community supports the project. (c) The more rewards for financial investments, the more likely will be that the money will flow.

Then everybody can and driven by the motivations will focus on the lowest hanging fruits first, then leverage token earnings to spread out benefits across other parts of society. Token rewards and token appreciation replace politics-driven tax-and-spend with proceeds based on actual behavior of people.

Some obvious challenges remain, including the link to fiat currencies or other coins. However, the issue may not be as severe as it seems. Even today people use different tokens, from fiat currency to phone minutes, foreign currencies, and different ones of all of these when doing business across borders. A likely answer will be to use a general and "hard" currency like the one previously suggested as a pan-African Fintech and commercial solution.

There is much more to it. A token enables comparisons and measurements. This also drives the establishment of rules for distributing revenues and can function as the basis for diverse governance. Decisions, e.g. on prices or other key measures, can be made based on coin ownership.

It enables the alignment of interests and creates a communal effort driving continuous infrastructure development and maintenance, and its profitable use. As an additional result, it also can establish and maximize long-term revenue streams for communities adjacent to such infrastructure projects. It is one more example of how a Blockchain-based token mechanism can accumulate value over time and let the owners and the community benefit from it.

Building transportation infrastructure becomes part of a virtuous cycle. Growth would be unlikely if nothing else were to change. However, this is how African heroes can use Blockchain to fulfill their destiny, by forging ahead in all areas of infrastructure. Then, steps like building roads, trains, airports, or ports will become profitable since they all reinforce building future-oriented value. Infrastructure investments become powerful when combined with the full range of investments in all core areas. These are the "1+5:" Fintech and commerce, communications and computing, education, health care, energy, and transport.

Starting with city-based but Africa-wide distributed boost centers, it then lays the foundation to continue to drive the creation of value across the whole economy, by producing raw materials, products, and services in all sectors. Agriculture, mining, trade, retail, construction, manufacturing, and much more will all benefit from it.

The combination of small steps in these areas will add up and enforce each other, driving aggressive economic growth. They all are future-oriented and profitable. This makes everything sustainable and self-reinforcing. Blockchain-technologies and related economic motivations and governance principles can turn them all into community-based efforts, with their vertical and horizontal dimension designed to ensure that everybody in society wins.

Africa's Heroes win, African brains win, Africa wins. The world wins since another billion has joined in, building, contributing, and improving everybody's future.

Epilogue: The First Trillionaires

DONALD TRUMP MAY NOT LIKE IT, but although Amazon's founder Jeff Bezos has much less hair, he has more money than him. A LOT more money even, making him possibly somewhere between 30 and 50 *times* as rich as the real estate developer and U.S. president.[194]

Far beyond such numbers, widespread Bitcoin or Ethereum adoption that turns them into "the world's money" would most likely result in somebody becoming 10 times as rich as Jeff Bezos is today.

It would create **the world's first trillionaire**.

From Billionaire to Trillionaire

The logic is simple. Even today, several individuals have become hundredfold millionaires, billionaires, or even multi-billionaires,[195] [196] by taking advantage of their unique position, knowledge, and involvement in the early stages of cryptocurrency development when Bitcoin and a few others of the largest Altcoins were launched.[197] Purely by sitting tight during a massive adoption of one of these cryptocurrencies, they could see a further explosion of their personal net asset value.

This may be acceptable if it results in pulling hundreds of millions, or billions, of people out of poverty and letting them pursue their own dreams in life. Or it may be anathema as a matter of principle. In any way this adds an important perspective to the pursuit of *"money for Africa."* Today's intellectual whiz kids, bold crypto-entrepreneurs, and highly invested VCs want you to buy in and use their kind of Blockchain-based coins.

Maybe they are above all altruistic, though, and so I am withholding judgment on motivations. But this still leaves us with a big question.

Which currency should underlie African financial transactions? Should one continue to focus on the US Dollar? The Euro? Create a new fiat-currency? Or simply leverage Bitcoin, as BTC or BCH? Use another cryptocurrency, like Ethereum, EOS, Stellar, Cardano, or even something new? Or should Africa create its own cryptocurrency?

This decision is not urgent (yet), because it first requires the solving of some open technical and economic issues, which then need to bear themselves out in the markets and in real life. However, we do not want to end up at an unfortunate place by default, just because we did not think things through enough. A few parameters are therefore important to consider.

An "African" currency can hardly be a fiat currency since this would mean a solution very similar to those already in existence today. Banks and intermediaries, influenced or even ruled by politicians, would control how much money would be created, who would create it, and who would "own" it. As a reminder, in today's central bank systems it is typically private banks that control how much money is created. In some systems, politicians provide some direct or indirect "guidance." But banks electronically create (most of) the money, and since it exists mainly as electronic records in their databases, they effectively "own" it.

And if they make mistakes, they get bailed out. Like in Vegas, the "house" always wins.

Even if this could be resolved, and banks, politicians, and intermediaries would relinquish control and their special benefits, the term "fiat" implies that trust in such a currency would be mandated top down. No more, but arguably also no less than 55 of Africa's 55 countries tried that approach, with little success. There needs to be more to it than simply telling people to use such artificial "fiat" money.

Above all, money is little more than a means of accounting for underlying value. It therefore needs to be linked to the value that has been or is being generated in a society, by its economy. As such its value is not as much "fiat" (mandated) but if anything, then much more "hard" or "de facto" (based on something real). Money supply and creation are then ideally more of a mechanical, or mathematical, challenge than a political one. Stable money requires a link to underlying value. Scarcity is good, but not enough.

If one would then link such a "stable" currency's supply and distribution to a mathematical formula, then one would not have a fiat currency anymore. Rather this would re-create something much equivalent to today's cryptocurrencies. The rest are technical details. They are challenging, but not new concepts.

So maybe a Blockchain-based currency sounds like a better option after all.

It could then be a cryptocurrency like Bitcoin, and if so as of the current state of affairs more likely a version like Bitcoin Cash (BCH) due to performance reasons and in line with currently expressed visions.[198] Ethereum still has many of the same limitations as BTC, while much more scalable solutions like EOS, Stellar, Cardano, and others can be challenged for not being decentralized enough or not having sovereign-level security.

And there we are back: This then poses the above trillionaire problem.

We may possibly have a governance issue also. Maybe it is just me, but hard forks and soft forks initiated and eventually executed by some techies and managers of crypto-mining companies do not instill much more confidence than decisions by the board of a Federal Bank, if even that. It involves personalities and economic interests here and there, and lots of technical and economic talk that ultimately is theory until users get hit with the consequences, and markets sort out "reality."

In principle, I am therefore suggesting to create and use a token linked to the value that African and other brains generate. This token may be tethered to one or a multitude of the above currencies, although maybe only initially. We should then let the markets guide us. Eventually, the token may turn into its own currency, with global use. Succeeding, it may become the currency for the Internet of Brains that I have above sketched out at a high level.

If implemented smartly, one huge advantage of such a currency is it not being distorted by the huge ownership imbalance inherent in all of today's cryptocurrencies. There will never be complete equality, and since humans and their aspirations differ, there never should be. But there also needs to be some balance.

Alternative scenarios are also possible. One of today's fiat currencies (the U.S. Dollar, Euro, Yen, Swiss Franc) may morph into more of a mathematically controlled one, and approach a new, market and community-based governance model similar to a cryptocurrency. Sure, bankers and politicians will give up power... And I will find some vibranium. Both could happen, theoretically.

Or the logic underlying today's mining protocols could be modified even much more profoundly than yet envisioned by some leaders of the Ethereum community, for example. New ETH coins could be linked to new economic value created, significantly diluting today's ownership value and even the power of mining pools.

The before-mentioned "Fab 5" of Ethereum have frequently mentioned that they do not like the establishment of a "plutocracy." For what it's worth, I believe them. It is their ability to deliver that I am more concerned about. They are very active and working on a shift from "Proof of Work" to "Proof of Stake," combined with "sharding" and other technical solutions.[199] This may solve the scalability issues or not – or these solutions growing complexities may have unpleasant unintended consequences.

Then why not start with something new after all? It may not be an "African Blockchain," as if mathematics could really be linked to geography. But like humanity, it may originate there, and become the most realistic money for Africa.

And it is not all bad that we then also could delay our meeting the first trillionaire.

Not seeing a trillionaire any time soon may be little consolation to most people, though, and particularly the middle class and the poor. Time is not on their side.

It's Now or Never: The Last Decade

"Once the masses lose their economic and political power, inequality levels could spiral alarmingly," claims Yuval Noah Harari, the author of the bestselling history and philosophy books *Sapiens*, *Homo Deus*, and *21 Lessons for the 21st Century*. He warns of a widening divide and the emergence of the most unequal societies that ever existed.[200]

This is a real danger for people in Africa. If he is right, almost all Africans may be left behind. Not despite technology, but *because* of technology, the gap can widen. As I tried to explain, I think it is even probable that the gap will widen, unless we see much bolder steps than trying to "catch up" over decades. The heroes of Africa must leverage emerging technologies to let, as fast as possible, most or all Africans join the world of the near future. They need to be able to reap the related economic benefits before the divergence gets too big and it becomes too late.

Africa's heroes need to create expertise and infrastructure across their continent. By using new crypto-economic motivations and structures they have to do so in a way that is scalable and efficient, not limited, controlled, and corrupt. By embracing and using emerging technologies they can establish a future-oriented commercial and financial infrastructure. This then enables the pursuit of the only realistic way forward, basing the winning strategy on the most important condition for such advancement and creation of wealth: human brains.

It is a strong argument for aggressively pursuing the vertical dimension of the strategy I outlined above. But according to Yuval Harari it may not be enough. In the same article quoted above, he even uses the same terms of "vertical" and "horizontal" to warn that "[g]lobalization will unite the world on a vertical axis and abolish national differences, but it will simultaneously divide humanity on a horizontal axis."[201]

African societies, therefore, need to apply new economic models and ways of thinking to spread the wealth, *rapidly and effectively*, to today's 800 million young Africans, and to their parents and children.

Centralized governmental solutions cannot reasonably be expected to achieve this fast enough. There is no precedence for the speed required. And traditional approaches are challenged and under siege even in the richest societies on Earth.

The key to success is decentralized and organically growing solutions based on crypto-technologies, like Blockchains. They can be combined with emerging technologies in communications and computing, biotech and particularly genomics, and renewable sources and storage of energy, to trigger massive investments into basic infrastructure. A rapidly increasing size of African economies will stimulate and enable entrepreneurial energy and startups that address such challenges of African and global societies.

But importantly, these skills and the inflow of for-profit private investments need to be established quickly. Advances in AI and computing may not give another chance. The window for gaining momentum could close in a mere decade. Once the rest of the world has designed, built, and stepped into the emerging technology-induced new world of the near future, everybody else may be left behind.

The heroes of Africa cannot let this happen. They can and have to play a leadership role in implementing forward-looking and radically successful solutions that advance all of humanity. This book, although not perfect and not comprehensive and detailed enough, outlined a plan that the heroes can execute.

I am under no illusions that there is a good chance that the above-listed obstacles may prove too strong. Blockchain and crypto-technologies may not be able to remove enough of Africa's constraints, controls, and corruption after all. People sometimes call it selfishness versus altruistic or social motivations.

Logically and conceptually, though, I trust that a decentralized Blockchain-focus can bridge that gap. My suggestions focus on private investments, even banking on each individual pursuing what is best for them. Only then solutions really are sustainable. Tokens and built-in crypto-economic motivations can help us align these interests to create a common good. If then somebody gets bailed out, everybody else needs to get bailed out, too.

The special circumstances of Africa make it the most fertile breeding ground for such solutions, and success will more positively affect a larger number of people than anywhere else.

But as I started in the opening sentence, nothing I am writing here is limited to Africa. It is about the potential of technology to transform and advance human societies anywhere.

I just hope that our African heroes and brains don't let the rest of the world try to figure things out by themselves.

Appendix

About the Author

Bert Kastel is a globally active entrepreneur who extensively works with Blockchain and other emerging technologies. His formal education is in economics, political science, business, and international relations.

Bert writes about "techtonic" shifts that transform societies and geopolitics, addressing profound questions of poverty and prosperity, war and peace.

Techies, Trust, and Trillionaires summarizes his insights of an intense six-month intellectual "journey" that started with him speaking at meet-ups in Nairobi and at the first ever Ethiopian Blockchain event in Addis Ababa in December 2017.

It culminated in early May 2018 with Bert co-organizing the by then biggest and most substantive Blockchain event in sub-Saharan Africa – *TheWorkshop* and *Blockchain Hub* at the Transform Africa Summit 2018 in Kigali.

Acronyms

3-D	Three-Dimensional
5G	5th Generation (Data Communications Standard)
ABC	Africa Boost Center
ADA	Cardano (Ticker Symbol)
AI	Artificial Intelligence
AIDS	Auto-Immune Deficiency Syndrome
AR	Augmented Reality
ASEAN	Association of Southeast Asian Nations
AU	African Union
BCH	Bitcoin Cash (Ticker Symbol)
BSIC	Blockchain for Social Impact Coalition (company)
BTC	Bitcoin (also "Bitcoin Core," Ticker Symbol)
CEO	Chief Executive Officer
CFTA	Continental Free Trade Area
CRISPR	Clustered Regularly Interspaced Short Palindromic Repeats
DIP	Digitized Individualized Production
DNA	DeoxyriboNucleic Acid
DPOS	Delegated Proof of Stake
DRC	Democratic Republic of Congo
EAMU	East African Monetary Union
EEF	Ethereum Enterprise Foundation
EKG	Electrocardiogram
EOS	EOS Blockchain (Ticker Symbol)
ETH	Ethereum (Ticker Symbol)
EU	European Union
Fed	Federal Bank of the United States of America

FOMO	Fear of Missing Out
GDP	Gross Domestic Product
GEO	Geostationary Earth Orbit
HEO	High-Earth Orbit
HODL	Hold On for Dear Life (Blockchain investment strategy)
HTS	High Throughput Satellites
ICO	Initial Coin Offering
IMF	International Monetary Fund (part of the UN)
IoB	Internet of Brains (also "IOB")
IOHK	Input Output Hong Kong (company)
IoT	Internet of Things
IOTA	Blockchain Company Name
IPFS	InterPlanetary File System (Blockchain-based decentralized storage technology)
LEO	Low-Earth Orbit
MEO	Medium-Earth Orbit
MOAP	Mother Of All Pipes (Satellites)
MIDS	Mediators of Individual Data
MIT	Massachusetts Institute of Technology
NFC	Near-Field Communication
NGO	Non-Governmental Organization
P2P	Peer-to-Peer
POS	Proof of Stake (also PoS)
POW	Proof of Work (also PoW)
R&D	Research and Development
RFID	Radio-Frequency Identification
ROI	Return on Investment
SIA	Company name (Blockchain, distributed storage)
TAS	Transform Africa Summit
Tbps	Terabites per second, Terabytes per second (trillion bites/bytes per second)
TR	Technology Review (MIT Magazine)
UN	United Nations
U.S.	United States (of America), also "US"

USA	United States of America
VC	Venture Capital
VLEO	Very-Low Earth Orbit
VR	Virtual Reality
WHO	World Health Organization
XLM	Stellar (Ticker Symbol)
XR	Extended Reality
XRP	Ripple (Ticker Symbol)

Blockchain Terms

Altcoin - Digital cryptocurrency similar to Bitcoin ("alternative to Bitcoin")

Bitcoin - First cryptocurrency, uses Blockchain technology to create and process digital cash

Blockchain - A digital public ledger to record cryptocurrency transactions

Blue Ocean - Also "Blue Ocean Strategy." Creation of a new, uncontested market space that makes competitors irrelevant and creates new consumer value.

Coin - See also: token. Units of a digital asset used as a medium of account and exchange.

Consensus Mechanism - Mechanism used in computer and Blockchain systems to achieve agreement on the state of a decentralized operating network.

Crypto-Economics - (1) A toolbox to determine the security of information in a system. (2) Economic motivations built into computer code to drive the behavior of actors in a network.

Cryptocurrency - Digital asset that can be a medium of account and exchange.

Delegated Proof of Stake (DPoS) - A consensus model that lets owners of a crypto-token delegate their voting rights to such persons that perform critical transactions for the maintenance of the Blockchain (like validating or confirming transactions, agreeing on the contents of blocks).

Feature Phone - A simple phone that can access the Internet but lacks advanced capabilities of a smart phone.

Fiat Money - Money made legal tender by decree of a government

Fintech - Financial Technology

Genesis Block - The first block of a Blockchain, usually hardcoded, including the ownership of coins or tokens that have been created.

Governance - Mechanisms used to define and enact the laws, rules, and norms of social interaction in a group

Leapfrogging - The concept of using radical innovation to bypass traditional incrementally progressing steps and reach or pass leading or dominant actors in a market.

Micro-Grid - A small network of producers, distributors, and users of electricity that operate independently of a centralized national electricity network.

mPesa - Mobile (m) Money (Pesa, Suaheli), a mobile phone-based money transfer and financing service, in Kenya and Tanzania

Permissioned - A type of Blockchain that requires identification and formal permission for performing certain actions.

Platform - A group of technologies (or applications) used as a base that other applications, technologies. or processes run on.

Proof of Stake (PoS) - The concept stating that important maintenance tasks of a Blockchain network, like creating new coins or validating blocks, can only be performed by owners of that coin, or even a subset of owners.

Proof of Work (PoW) - The concept requiring a not insignificant amount of effort (with Bitcoin or Ethereum: energy-intense calculations to solve a mathematical riddle) to perform important tasks of a Blockchain network, like creating new coins or validating blocks. Since this effort involves cost, it is a key tool to keep work "honest."

Sharding - Breaking up of a database into separate pieces that are able to some degree to operate independent of each other.

Token - See also: coin. The unit of a digital asset that reflects value and can account for, store, and transfer that value. Formally, the difference to a crypto coin is that a token has a "utility" and is not purely an abstract unit of a currency. In reality, coin and token are almost identical.

Trilemma - A design challenge introduced because only two of the following three are considered achievable at the same time:

scalability/speed, security, and decentralization

Trustless - The functioning of a social network in a way that one does not need to trust other actors or middlemen, instead enforcing "honest" behavior of all participants through mathematical algorithms and economic motivators.

Companies

Blockchain for Social Impact Coalition. Incubator, developer, and implementer of Blockchain products that can address social and environmental challenges

Better World Initiatives Africa. Nairobi-based startup focusing on innovation and development

Bill & Melinda Gates Foundation. Foundation focusing on work that helps people living healthy and productive lives

BitPesa. African Blockchain startup focusing on fast, easy, and low-cost money transfers

Bitcoin.com. Bitcoin (BTC and BCH) miner and cryptocurrency services company

Block.one. Blockchain company that incubated EOS

Cardano. Blockchain technology and platform focusing on peer-reviewed and science-based advances

Chamapesa. Nairobi Blockchain startup, focusing on communal identity

Consensys. Venture Production Studio focusing on Ethereum

EOS. Blockchain, founded by Dan Larimer

Eversend. Blockchain and Fintech startup from Uganda

Filecoin. Blockchain company focusing on distributed storage

Ford Foundation. Foundation focusing on social justice

Gebeya. Ethiopian technology company and incubator (Gebeya is Amharic for "Marketplace")

Healid. Blockchain project focusing on collecting and analyzing the human genome through a decentralized platform

Hyperledger. Open source project of the Linux Foundation to advance cross-industry Blockchain technologies

Icog Labs. Ethiopian AI company, known for humanoid robot

Sophia

IOHK. Blockchain company, founded by Charles Hoskinson

IOTA. Crypto-technology company, known for their "tangle"

KiWi New Energy. Blockchain startup in energy

kLab. Rwandan incubator and innovation hub

Kukulu. Ethiopian 3D mobile game, founded by Dawit Abraham, incubated by Gebeya

MIT. Massachusetts Institute of Technology, University

nChain. Blockchain research and development company, focusing on Bitcoin Cash (BCH)

Pesanest. Ugandan Blockchain startup and micro-lending platform (now called Quzacredit)

Protocol Labs. Blockchain company, among others the creators of Filecoin

Quzacredit. Ugandan Blockchain startup and micro-lending platform (formerly Pesanest)

Ripple. Blockchain company (permissioned)

Rwanda Central Bank. Central Bank of the nation of Rwanda

SIA. Blockchain company focusing on distributed storage

SingularityNET. Blockchain company establishing a decentralized AI network

Soramitsu. Japanese Blockchain technology company, known to be the creator of Hyperledger Iroha

Stellar. Blockchain focusing on the future of banking

Tech844. Company enabling emerging technologies

UWINCorp. Venture Capital and Incubation company for Blockchain ventures

Wala. African Blockchain startup in Fintech, focusing on zero-fee money transfers

WePower. Blockchain startup, green energy trading platform

Chan Zuckerberg Initiative. Foundation aiming to advance human potential and equal opportunity, with an initial focus on biomedical research and education

People

Aanyu, Rachael - Founder of Pesanest and Quzacredit

Abraham, Dawit - Creator of Kukulu, a 3D mobile game out of Ethiopia

Arbesman, Samuel - Author of *The Half-Life of Facts*

Berhane, Natnael - Student and Blockchain developer from Mauritius

Bezos, Jeff - Founder and CEO of Amazon

Bingabo, Maximillien - Programmer and Blockchain developer from Rwanda

Buterin, Vitalik - Founder of Ethereum

Daffe, Amadou - Founder of Gebeya

Diamond, Jared - Author of Pulitzer-winning *Guns, Germs, and Steel*

Drake, Justin - Researcher at the Ethereum Foundation

Durban, Dick - U.S. Senator

Floersch, Karl - Ethereum Casper Core Developer

Francois - Printer in Kigali

Gates, Bill - Founder of Microsoft, Co-CEO of the Bill & Melinda Gates Foundation

Gates, Melinda - Co-Founder of the Bill & Melinda Gates Foundation

Gichinga, Njoki - Manager for partnership at Safaricom Innovation Center (and formerly iHub)

Goertzel, Ben - Founder of SingularityNET, and Icog Labs in Ethiopia)

Grigg, Ian - Founder of Chamapesa (and well-known cryptographer, involved with Block.one and EOS, among many others)

Haguma, Norbert - Co-Founder AfricaGen, Head of Blockchain Hub at Smart Africa

Harari, Yuval Noah - Author of *Sapiens, Homo Deus, 21 Lessons for the 21st Century*

Hoskinson, Charles - Technology entrepreneur and mathematician, founder of IOHK and of the Cardano Blockchain

Japheth, CK - Startup Ecosystem Builder at The Innovation Village Kampala

Joab - Uber driver

Kagame, Paul - President of the Republic of Rwanda

Kastel, Fiona - Student of the University of Utah, supporting the organization of a Blockchain workshop in Kigali

Kajangwe, Alain - Co-Founder and CEO of WiredIn (Rwanda)

Kenyatta, Uhuru - President of the Republic of Kenya

Kiconco, Olga - Founder of Pesanest and Quzacredit

Ladipo, Tunde - Director of Partnerships at Interstellar (Stellar)

Lan, Cui Wu - Chinese student (formerly)

Lanier, Jaron - Computer Scientist, Writer, considered the founder of the field of Virtual Reality

Larimer, Dan - Founder of BitShares, Steemit, and EOS, also CTO of Block.com

Lemma, Markos - CEO and co-Founder of iceaddis innovation hub

Ma, Jack - Founder and CEO of Alibaba

Martyniuk, Nikolay - Co-Founder of WePower, a Blockchain startup in energy

Miyaguchi, Aya - Executive Director, Ethereum Foundation

Musk, Elon - Founder and CEO of SpaceX and Tesla

Mutangana, Aphrodice - General Manager of kLab (Rwandan Incubator and Innovation Center)

Owen, Cameron - Student of the University of Utah, supporting the organization of a Blockchain workshop in Kigali

Piik, Helinä - Founder of Better World Initiatives Africa

Poon, Joseph - Founder of Lightning Network, Ethereum Core Developer

Raju, K Rama - Managing Director, Sqanda Group

Ro, Sandra - CEO at Global Blockchain Business Council
Shin, Dongyeol - Manager at Samsung SDS
Soohyoung, Yoon - General Manager at J&C Corporation, formerly manager at Samsung Engineering (SECL)
Sun, Irene - Kiwi New Energy Marketing Communications Manager
Takemiya, Makoto - Co-CEO at Soramitsu
Taleb, Nassim - Author of *The Black Swan*, *Antifragile*, and *Skin in the Game*; former trader and risk analyst
Thatcher, Margaret - British Prime Minister from 1979 to 1990
Trump, Donald - U.S. President since 2016; Real Estate Developer and Entertainer
Ver, Roger - CEO of Bitcoin.com and early investor in Bitcoin-related startups
Weyl, Glen - Author of *Radical Markets*, Principal Researcher at Microsoft, Economist
Wright, Craig - Chief Scientist at nChain, Computer Scientist and Blockchain entrepreneur
Xi, Jinping - President of the People's Republic of China
Yew, Lee Kuan - Founder of the sovereign nation of Singapore
Zamfir, Vlad - Ethereum core developer

List of Figures

Index

Notes

1 https://www.ethereum.org/foundation

2 https://deconomy.com

3 Sarah Boumphrey, Special Report: The World's Youngest Populations (Euromonitor International, 13 Feb 2012), https://blog.euromonitor.com/2012/02/special-report-the-worlds-youngest-populations-.html [7 Oct 2018]

4 Kenyan WallStreet Editorial, *Kenya among the few countries in the world with highest per capita holding in Bitcoin – Citi* (The Kenyan Wall Street, 12 Jan 2018) https://kenyanwallstreet.com/kenya-among-countries-world-highest-per-capita-holding-bitcoin-citi/ [7 Oct 2018]

5 Roger made this and the following statements during his debate with Blockstream's CEO Samson Mow, at Deconomy 2018 on April 3, 2018. See https://www.youtube.com/watch?v=zSbPz4g9rZQ

6 Ibid. With this specific statement he was directly addressing Blockstream and Bitcoin, although its contents can be applied to a much larger portion of the Blockchain community.

7 https://www.getwala.com

8 https://public.bitpesa.co

9 https://chamapesa.com

10 https://www.eversend.co

11 https://new.consensys.net

12 https://www.blockchainforsocialimpact.com

13 https://nchain.com/en/

14 https://www.cardano.org/en/home/

15 https://www.stellar.org

16 Soramitsu is the main developer of Hyperledger Iroha, a permissioned Blockchain technology. See https://soramitsu.co.jp/en/

17 CryptoCurrency, *'I've Got More Money You're your Country': Craig Wright tells Rwanda* (CryptoCurrency, 25 May 2018) http://cryptotimes.org/cryptocurrency/ive-got-more-money-country-craig-wright-tells-rwanda/ [7 Oct 2018]

18 Nassim Nicholas Taleb, *Skin in the Game: Hidden Asymmetries in Daily Life* (New York: Random House 2018)

19 Unless explicitly stated differently, all currency numbers mentioned here are in US Dollars.

20 CryptoCurrency, *'I've Got More Money You're your Country': Craig Wright tells Rwanda* (CryptoCurrency, 25 May 2018) http://cryptotimes.org/cryptocurrency/ive-got-more-money-country-craig-wright-tells-rwanda/ [7 Oct 2018]

21 Salvatore Babones, *India Will Outgrow China In 2018 But Must Invest In Next-Generation Value Chains To Succeed* (Forbes, 2 Jan 2018) https://www.forbes.com/sites/salvatorebabones/2018/01/02/india-will-outgrow-china-in-2018-but-must-invest-in-next-generation-value-chains/#5927ff8d44d8 [7 Oct 2018]

22 John Mbu, *Despite headwinds, Africa's economy is still strong* (World Economic Forum, 29 Jul 2016) https://www.weforum.org/agenda/2016/07/african-economic-situation-the-fundamentals-still-remain-strong/ [7 Oct 2018]

23 Acha Leke, *3 reasons thinks are looking up for African economies* (World Economy Forum, 5 May 2016) https://www.weforum.org/agenda/2016/05/what-s-the-future-of-economic-growth-in-africa/ [7 Oct 2018]

24 http://www.worldometers.info/world-population/india-population/ [10 Oct 2018]

25 https://en.wikipedia.org/wiki/Languages_of_Africa [10 Oct 2018]

26 https://en.wikipedia.org/wiki/Languages_of_India [10 Oct 2018]

27 Depending on the source, the numbers vary widely, particularly for Africa, between US$2.3 billion and US$3.3 billion. The

numbers here are according to the International Monetary Fund (IMF), *World Economic Outlook (October 2018)* [10 Oct 2018] qhttps://www.imf.org/external/datamapper/NGDPD@WEO/OEMDC/ADVEC/WEOWORLD

[28] Tom Holland, *The Risk of the Middle-Income Trap Just Increased For China. Here's Why.* (South China Morning Post, 23 Oct 2017) http://www.scmp.com/week-asia/opinion/article/2116323/risk-middle-income-trap-just-increased-china-heres-why [7 Oct 2018]

[29] Alyssa Ayres, *Can Indian Manufacturing Be the Next Chinese Manufacturing?* (The Atlantic, 4 Jan 2018) https://www.theatlantic.com/business/archive/2018/01/india-manufacturing-tata-alyssa-ayres/549263/ [7 Oct 2018]

[30] Prosenjit Datta, *India must become globally competitive in manufacturing to boost merchandise exports* (Business Today, 5 Jun 2018) https://www.businesstoday.in/opinion/prosaic-view/india-must-become-globally-competitive-in-manufacturing-to-boost-merchandise-exports/story/276100.html [7 Oct 2018]

[31] Dan Gilmore, *Supply Chain Predictions for 2018 from Gartner and IDC* (SupplyChainDigest, 2 Mar 2018) http://www.scdigest.com/firstthoughts/18-03-02.php?cid=13810 [22 Oct 2018]

[32] Gina Roos, *IDC's 10 Supply Chain Predictions for 2018* (EPSNews, 13 Dec 2017) https://epsnews.com/2017/12/13/idcs-10-supply-chain-predictions-2018/ [22 Oct 2018]

[33] Ibid.

[34] PTI, *Achieving 25% of GDP from manufacturing, 100 million jobs by 2022 difficult: Survey* (The Economic Times, 27 Sep 2015) https://economictimes.indiatimes.com/news/economy/indicators/achieving-25-of-gdp-from-manufacturing-100-million-jobs-by-2022-difficult-survey/articleshow/49124619.cms [7 Oct 2018]

[35] Hernan Galperin, Andrea Alarcon, *The Future of Work in the Global South* (Ottawa, Canada: IDRC – International Development Research Center, 2017? [no date])

http://fowigs.net/future-work-global-south-2/ [22 Oct 2018]

36 The term "Blue Ocean Strategy" was popularized by the book of the same name. W. Chan Kim, Renée Mauborgne, *Blue Ocean Strategy* (Boston: Harvard Business Review Press, 2015)

37 Curtis Abraham, *Interview: The father of the internet in Africa* (NewScientist, 6 Aug 2008)
https://www.newscientist.com/article/mg19926681-900-interview-the-father-of-the-internet-in-africa/ [22 Oct 2018]

38 https://www.internethalloffame.org/inductees/nii-quaynor

39 AfricaHYS Team, *Should Africa have fewer babies?* (BBC, 10 May 2011)
http://www.bbc.co.uk/blogs/africahaveyoursay/2011/05/should-africa-have-fewer-babie.shtml [7 Oct 2018]

40 The European Commission claims that "27% of the African population would like to move to another country." European Commission, *New perspectives on African migration* (European Commission, 1 Jun 2018)
https://ec.europa.eu/jrc/en/news/new-perspectives-african-migration [12 Oct 2018]

41 *This Week*, Thames TV, 5 Feb 1976
http://www.oxfordreference.com/view/10.1093/acref/9780191826719.001.0001/q-oro-ed4-00010826

42 Herein, and to keep it simple, I use the terms "tokens" and "coins" interchangeably.

43 https://kioneki.com/michael-kimani/

44 Duncan J. Watts, *Six Degrees – The Science of a Connected Age* (New York: W. W. Norton & Company 2003, 2004)
https://www.amazon.com/Six-Degrees-Science-Connected-Age/dp/0393325423

45 Stefan Schrauf, Philipp Berttram, *Industry 4.0: How digitization makes the supply chain more efficient, agile, and customer-focused* (Strategy&, 7 Sep 2016)
https://www.strategyand.pwc.com/reports/digitization-more-efficient [22 Oct 2018]

46 Using a different kind of calculation, McKinsey comes to a similar dimension of transformative potential, pointing at a transformative potential of more than $1 trillion for the *Indian* economy, focusing on 12 technologies in the areas of "digitized

life and work," "smart physical systems," and "energy." See: Noshir Kaka, Anu Madgavkar, James Manyika, Jacquest Bughin, Pradeep Parameswaran, *India's tech opportunity: Transforming work, empowering people* (McKinsey Global Institute, Dec 2014) https://www.mckinsey.com/industries/high-tech/our-insights/indias-tech-opportunity-transforming-work-empowering-people [22 Oct 2018]

47 Gil Press, *Top 10 Tech Predictions For 2017 From IDC* (Forbes, 1 Nov 2016) https://www.forbes.com/sites/gilpress/2016/11/01/top-10-tech-predictions-for-2017-from-idc/#19a5865b4aad [22 Oct 2018]

48 *Speech by Mr. Lee Kuan Yew, Senior Minister, for Africa Leadership Forum at the Regency Hotel on Monday, 8. November 1993* (National Archives of Singapore, 1993) http://www.nas.gov.sg/archivesonline/data/pdfdoc/lky19931108.pdf [7 Oct 2018]

49 Adam Whithnall, *Is the entire GDP of Africa really just equal to that of France?* (Independent, 18 Oct 2016) https://www.independent.co.uk/news/world/africa/is-the-entire-gdp-of-africa-really-just-equal-to-that-of-france-a7367906.html

50 Mark J. Perry, Economic Output: *If States Were Countries, California Would Be France* (Newsweek, 11 June 2016) http://www.newsweek.com/economic-output-if-states-were-countries-california-would-be-france-467614 [7 Oct 2018]

51 Robbie Gramer, *Infographic: Here's How the Global GDP Is Divvied Up* (Foreign Policy, 24 Feb 2017) http://foreignpolicy.com/2017/02/24/infographic-heres-how-the-global-gdp-is-divvied-up/ [7 Oct 2018]

52 Calculated as 1.2 billion of 7.5 billion overall.

53 This calculation assumed 1.2 billion Africans and 7.5 billion people on Earth, based on UNESCO, Statistics on Youth (UNESCO, 13 Feb 2013) http://www.unesco.org/new/en/unesco/events/prizes-and-celebrations/celebrations/international-days/world-radio-day-2013/statistics-on-youth/ [7 Oct 2018]

54 Assuming about 70% of the population in Africa, with the

comparison based on Sarah Boumphrey, *Special Report: The World's Youngest Populations* (Euromonitor, 13 Feb 2012) https://blog.euromonitor.com/2012/02/special-report-the-worlds-youngest-populations.html [7 Oct 2018]

55 James Manyika, *10 imperatives for Europe in the age of AI and automation* (McKinsey Global Institute, October 2017) https://www.mckinsey.com/featured-insights/europe/ten-imperatives-for-europe-in-the-age-of-ai-and-automation [22 Oct 2018]

56 Vijai Govindarajan, Gunjan Bagla, *Understanding the Rise of Manufacturing in India* (Harvard Business Review, 18 Sep 1015) https://hbr.org/2015/09/understanding-the-rise-of-manufacturing-in-india [7 Oct 2018]

57 John Paul Soltesz, Marni Rutkofsky, Karen Kerr, Marco Annunziata, *The workforce of the future – Advanced manufacturing's impact on the global economy* (GE, April 2016) https://www.gita.org.in/Attachments/Reports/AM_FINAL.pdf [7 Oct 2018]

58 This is not just true for space exploration, but a general trend. See Katie Brigham , Lora Kolodny, *This floating robotic factory will build satellites and spaceships in orbit* (CNBC, 23 Jun 2018) https://www.cnbc.com/2018/06/22/made-in-space-created-a-robotic-factory-to-build-satellites-in-orbit.html [7 Oct 2018]

59 Drew Desilver, *Most Americans unaware that as U.S. manufacturing jobs have disappeared, output has grown* (Pew Research, 25 Jul 2017) http://www.pewresearch.org/fact-tank/2017/07/25/most-americans-unaware-that-as-u-s-manufacturing-jobs-have-disappeared-output-has-grown/ [7 Oct 2018]

60 Fredrick Ngugi, *Uganda Shines with Highest Number of Women Entrepreneurs* (Face2Africa, 8 Mar 2017) https://face2faceafrica.com/article/women-entrepreneurs-uganda [7 Oct 2018]

61 https://www.urbandictionary.com/define.php?term=troll [7 Oct 2018]

62 https://www.merriam-webster.com/dictionary/trolling [7 Oct 2018]

63 Seema Jayachandran, *Think Positive, Climb Out of Poverty? It Just Might Work* (New York Times, 13 Jul 2018) https://www.nytimes.com/2018/07/13/business/think-positive-climb-out-of-poverty-it-just-might-work.html [21 Oct 2018]

64 Sibahle Malinga, *Tech hubs power funding to Africa's start-ups* (itweb, 23 Mar 2018) https://www.itweb.co.za/content/wbrpO7gPXBgMDLZn [11 Oct 2018]

65 Tony Tiyou, *Night Satellite Picture of Power Distribution across the globe* (Renewables in Africa, no date) https://www.renewablesinafrica.com/tony-tiyou/ [7 Oct 2018]

66 Rémi Jedwab, Edward Kerby, Alexander Moradi, *How colonial railroads defined Africa's economic geography* (Vox EU, 2 Mar 2017) https://voxeu.org/article/how-colonial-railroads-defined-africa-s-economic-geography [7 Oct 2018]

67 Michael Kende, *Promoting the African Internet Economy* (Internet Society, 22 Nov 2017) https://www.internetsociety.org/resources/doc/2017/africa-internet-economy/ [7 Oct 2018]

68 Samira Sadeque, *Smartphones lost market share to feature phones in Africa last year* (Quartz Africa, 13 Feb 2018) https://qz.com/africa/1206462/smartphones-lost-market-share-to-feature-phones-in-africa-last-year/ [7 Oct 2018]

69 Simon Kemp, *Digital in 2018: World's Internet users pass the 4 billion mark* (we are social, 30 Jan 2018) https://wearesocial.com/blog/2018/01/global-digital-report-2018 [7 Oct 2018]

70 Mariama Awumbila, *Drivers of Migration and Urbanization in Africa: Key Trends and Issues* (Center for Migration Studies, University of Ghana, Background Paper prepared for UN Expert Group Meeting on Sustainable Cities, Human Mobility and International Migration, 7-8 Sep 2017) http://www.un.org/en/development/desa/population/events/pdf/expert/27/papers/III/paper-Awunbila-final.pdf

71 Christina Sterbenz, *This One Country Shows How Distorted Our Modern Map Is* (Business Insider, 29 May 2014)

https://www.businessinsider.com/greenland-africa-comparison-2014-5 [7 Oct 2018]

72 Max Fisher, *A revealing map of the world's most and least ethnically diverse countries* (Washington Post, 16 May 2013) https://www.washingtonpost.com/news/worldviews/wp/2013/05/16/a-revealing-map-of-the-worlds-most-and-least-ethnically-diverse-countries/?utm_term=.a6c501a4e06f [7 Oct 2018]

73 This can be questioned when considering the substantial intra-African migrations of the past decades, including large number of refugees.

74 Other groupings are the Arab Maghreb Union (UMA), the Common Market for Eastern and Southern Africa (COMESA), the Community of Sahel-Saharan States (CEN-SAD), the East African Community (EAC), the Economic Community of Central African States (ECCAS), the Economic Community of West African States (ECOWAS), the Intergovernmental Authority on Development (IGAD), and the Southern African Development Community (SADC).

75 Mariama Sow, *Figures of the week: Africa's intra- and extra-regional trade* (Brookings, 29 Mar 2018) https://www.brookings.edu/blog/africa-in-focus/2018/03/29/figures-of-the-week-africas-intra-and-extra-regional-trade/ [7 Oct 2018]

76 For that reason this gets much closer to economic concepts of "pareto optimization" or even "total surplus maximization." In this context, and looked at in isolation, "pareto optimal" means that nobody would be better off by not following the blue ocean strategy, while there are lots of winners. Even more so, a blue ocean strategy as outlined here increases the overall value for society so significantly that it may be the best possible strategy of all for economic development, which is close to the definition of "total surplus maximization." Either one of these concepts establishes a superior result to a frequently targeted situation where everybody in society "settles" for an outcome, which in game theory is the essence of a "Nash equilibrium."

77 Jaron Laier, E, Glen Weyl, "A Blueprint For A Better Digital

Society" (Harvard Business Review, 26 Sep 2018)

78 I am not going into details for appropriate voting mechanism. Voting may be on-chain or off-chain, but probably involves some aspect of delegation, probably to the ABCs. New concepts of rapid voting in the form of a "liquid democracy," or even "quadratic voting" could and should be explored.

79 By it becoming a core component of the linking of global digitized supply chains, it lays the foundation for an even more future-oriented economic and social goal: linking the creators back to themselves. In an idealized world of 25 or more years in the future, all humans would be directly or indirectly part of such supply chains. Since they would in essence create value (the products) for other humans, in summarized form everybody would work for everybody. If everybody is compensated with coins that enable this process, built-in compensation mechanisms can ensure that the newly created monetary value is more equitably distributed.

80 Rani Molla, *Tech companies are spend more on R&E than any other companies in the U.S.* (recode, 1 Sep 2017) https://www.recode.net/2017/9/1/16236506/tech-amazon-apple-gdp-spending-productivity [11 Oct 2018]

81 Based on data from the CIA World Factbook 2016 and 2017, and summarized at: https://en.wikipedia.org/wiki/List_of_countries_by_government_budget

82 *Speech by Mr. Lee Kuan Yew, Senior Minister, for Africa Leadership Forum at the Regency Hotel on Monday, 8. November 1993*, page 21 (National Archives of Singapore, 1993) http://www.nas.gov.sg/archivesonline/data/pdfdoc/lky19931108.pdf

83 David Herbling, *Kenya Election Agency to Adopt Blockchain for Vote Transparency* (Bloomberg Quint, 20 August 2018) https://www.bloombergquint.com/business/2018/08/20/kenya-elections-agency-to-adopt-blockchain-for-vote-transparency [11 Oct 2018]

84 Lee Kuan Yew, Ibid.

85 The Bitcoin White Paper's rationale is built on countering the

"inherent weaknesses" of a system relying on financial institutions, and the first mined block contains the text "The Times 03/Jan/2009 Chancellor on brink of second bailout for banks."

86 Fiat money is artificial and not backed by anything tangible or intangible. People use it because a government says they have to. Once people stop following or believing, most or all of its value disappears. Examples for "fiat money" are US Dollars, Euro, Kenyan Shilling, or Argentinian Peso.

87 Only accepting a solution that has no losers is called "pareto optimal."

88 http://quzacredit.com

89 Yinka Adegoke, *Africa's startup funding deals are entering the million dollar era* (Quartz Africa, 29 Apr 2018) https://qz.com/africa/1264614/africas-venture-capital-deals-are-entering-the-1-million-era/ [7 Oct 2018]

90 http://www.gebeya.com

91 https://kukulugame.com/

92 Kamil Olufowobi, *Opinion: The 'door of return' is open for people of African descent* (CNN, 27 May 2017) https://www.cnn.com/2017/05/25/africa/africans-doors-of-return/index.html [7 Oct 2018]

93 For the U.S., by now far more Africans immigrated voluntarily than as slaves. See: Sam Roberts, *More African enter the U.S. Than in Days of Slavery* (New York Times, 21 Feb 2005) https://www.nytimes.com/2005/02/21/nyregion/more-africans-enter-us-than-in-days-of-slavery.html [7 Oct 2018]

94 Pew Research Center, *At Least a Million Sub-Saharan Africans Moved to Europe Since 2010* (Pew Research Center, 22 Mar 2018) http://www.pewglobal.org/2018/03/22/at-least-a-million-sub-saharan-africans-moved-to-europe-since-2010/ [7 Oct 2018]

95 United Nations, *International Migration Report 2017 Highlights*, p. 11 (New York: United Nations, 2017) http://www.un.org/en/development/desa/population/migration/publications/migrationreport/docs/MigrationReport2017_Highlights.pdf [12 Oct 2018]

96 According to Foundation Center, in 2012 there were about 2000 from the U.S. alone. See: Foundation Center, in cooperation with Africa Grandmakers' Affinity Group, *U.S. Foundation Funding for Africa – 2015 Edition* (Foundation Center, 2015) http://foundationcenter.org/gainknowledge/research/pdf/funding_for_africa2015.pdf [7 Oct 2018]

97 Ibid.

98 Ali Diallo, *Africa is the next frontier for VCs* (Venture Beat, 1 Apr 2017) https://venturebeat.com/2017/04/01/africa-is-the-next-frontier-for-vcs/ [7 Oct 2018]

99 Jared M. Diamond, *Guns, Germs, and Steel: The Fates of Human Societies* (New York: W.W. Norton & Company, 1999)

100 Blockchain News, $414 Billion Alibaba's Jack Ma: *Blockchain Needs to Target Manufacturing Industry* (CCN, 18 Sep 2018) https://www.ccn.com/414-billion-alibabas-jack-ma-blockchain-needs-to-target-manufacturing-industry/ [7 Oct 2018]

101 *Address by President Paul Kagame at 4th Transform Africa Summit* (Paul Kagame, 8 May 2018) http://paulkagame.com/?p=12811 [7 Oct 2018]

102 At the latest since the 19th Congress of the Chinese Communist Party (CCP) in 2017, President Xi's outlined China's new role in the world much more assertively. See: Tom Phillips, *Xi Jinping heralds 'new era' of Chinese power at Communist party congress* (The Guardian, 18 Oct 2017) https://www.theguardian.com/world/2017/oct/18/xi-jinping-speech-new-era-chinese-power-party-congress [7 Oct 2018]

103 He Qinglian, *How Africa Has Become a Second Front in the US-China Struggle* (The Epoch Times, 12 Sep 2018) https://www.theepochtimes.com/how-africa-has-become-a-second-front-in-the-us-china-struggle_2657439.html [7 Oct 2018]

104 Chaorong Wang, *Foreign Aid to Africa: The United States vs. China* (The Borgen Project, 23 Apr 2018) https://borgenproject.org/foreign-aid-to-africa-the-united-states-vs-china/ [7 Oct 2018]

105 OECD, *Development aid rises again in 2016* (Paris: OECD, 11 Apr 2017) http://www.oecd.org/dac/financing-sustainable-

development/development-finance-data/ODA-2016-detailed-summary.pdf [7 Oct 2018]

106 Press Release: *World Bank Group Announces Record $57 Billion for Sub-Saharan Africa* (World Bank, 19 Mar 2017) http://www.worldbank.org/en/news/press-release/2017/03/19/world-bank-group-announces-record-57-billion-for-sub-saharan-africa [7 Oct 2018]

107 Press Release: *World Bank Prices First Global Blockchain Bond, Raising A$110 Million* (Washington/Sydney, 23/24 Aug 2018) https://www.worldbank.org/en/news/press-release/2018/08/23/world-bank-prices-first-global-blockchain-bond-raising-a110-million [7 Oct 2018]

108 Sebastián Edwards, *How effective is foreign aid? (World Economic Forum*, 28 Nov 2014) https://www.weforum.org/agenda/2014/11/how-effective-is-foreign-aid/ [12 Oct 2018]

109 HODL is the acronym for "Hold On for Dear Life," indicating an investment strategy where one buys and holds on to the coin no matter what happens to the course in the short or even medium term.

110 "Smart contract" is a fancy word for conditional logic associated with a Blockchain transaction. For example, a payment may only be executed at a certain date, or when the receipt of a product has been confirmed.

111 Ji Soo Kim, Young Man Kwoun, Yowhan Son, Sue Kyoung Lee, *The history of deforestation and forest rehabilitation in Korea* (Korea University/University of Illinois at Urbana-Champaign, ca. 2011 [no date given]) https://www.academia.edu/4043937/The_history_of_deforestation_and_forest_rehabilitation_in_Korea

112 All numbers per The World Bank, *GDP per capita (current US$)*, according to World Bank national accounts data and OECD National Accounts data files. The corresponding numbers for the U.S. and U.K. are $3,827 and $1,851 for 1965 and $59,531 and $39,720 for 2017. See: https://data.worldbank.org/indicator/NY.GDP.PCAP.CD?contextual=default&end=1965&locations=KR-GH-JP-

SG&start=1965&view=bar [23 Oct 2018]

113 The Economist Briefing, *The battle of three centuries - The history of central banks* (The Economist, 27 Apr 2017) https://www.economist.com/briefing/2017/04/27/the-history-of-central-banks [8 Oct 2018]

114 Mary Bellis, *History of Roads in America and First Federal Highway* (30 Sep 2018) https://www.thoughtco.com/history-of-american-roads-4077442 [8 Oct 2018]

115 IRS, *The History of Income Taxes* (IRS) https://www.irs.com/articles/the-history-of-income-taxes [8 Oct 2018]

116 Jacob Funk Kierekegaard, *The True Levels of Government and Social Expenditures in Advanced Societies* (Peterson Institute for International Economics, Policy Brief Number PB15-4, March 2014) https://piie.com/publications/pb/pb15-4.pdf [19 Oct 2018]

117 Kate De Klerk, Mayra Salazar, *It's time for a single-payer health care system in the US. Here's why.* (The Do, 17 Jul 2018) https://thedo.osteopathic.org/2018/07/its-time-for-a-single-payer-health-care-system-in-the-us-heres-why/ [3 Nov 2018]

118 Drew Desilver, *What does the federal government spend your tax dollars on? Social security programs, mostly* (Pew Research Center, Facttank, 4 Apr 2017) http://www.pewresearch.org/fact-tank/2017/04/04/what-does-the-federal-government-spend-your-tax-dollars-on-social-insurance-programs-mostly/ [19 Oct 2018]

119 Blockchain News, *Cardano to Help Ethiopia Grow Coffee Using Blockchain Tech* (CCN, 6 May 2018) https://www.ccn.com/cardano-to-help-ethiopia-grow-coffee-using-blockchain-tech/ [8 Oct 2018]

120 Jayme Hoffman, *First Principles and the Art of Thinking like Elon Musk* (Medium: The Mission Podcast, 15 Sep 2016) https://medium.com/the-mission/first-principles-and-the-art-of-thinking-like-elon-musk-98658bb36569 [8 Oct 2018]

121 For arguments sake, let's assume the value generated to be US$50 per hour. Each person works for 2,000 hours a year, and 1% of Africa's population are at least 10 million people. $50*2,000*10 million = $1 trillion

[122] IPFS - InterPlanetary File System, a decentralized data storage technology created by *Protocol Labs*

[123] https://www.iota.org

[124] Per the Bitcoin White Paper. Ibid.

[125] The English system was way too narrowly focused on a small class of people, while the Swiss were much more informal, and both (but particularly the Swiss) covered a much smaller geographical region.

[126] Again, the term "Blockchain" is loosely defined here, and could for example also be called "self-organizing mathematically or cryptographically enforced networks."

[127] Have you tried to keep pests out of your home, in the summer?

[128] nChain's Craig Wright made the statement, and subsequent debates about the topic in general included Blockstream's Samson Mow, Bitcoin.com's Roger Ver, and Ethereum's Vitalik Buterin and Joseph Poon. See a video at https://www.youtube.com/watch?v=OFJgdhvNp54.

[129] In graph and network theory, nodes (also called "vertices") are connected using "edges" and the ratio between the total number of edges and the total number of vertices determines the network's centrality.

[130] They do so either for the whole network, or by breaking the network up into smaller pieces during the process of validating and reaching consensus.

[131] As reported on June 14, 2018 on https://dashboard.stellar.org

[132] As reported on June 14, 2018 on https://bchnodes.online, https://bitnodes.earn.com/?, https://www.ethernodes.org/network/1

[133] Proof of Work (PoW) is cryptographic approach that uses the search for the solution to a mathematical riddle to reach consensus on the correctness of a block. Proof of Stake (PoS) uses "staked" ownership of coins as a condition to validate the truth. "Sharding" breaks up a database into separate pieces that are able to some degree to operate independent of each other.

[134] This addresses the challenges related to the so-called "trilemma," stating that only two of the following three can be achieved at the same time: scalability/speed, security, and decentralization.

135 According to Wikipedia as of 8 Oct 2018, Africa has 55 nations, Asia 50, Europe 46, the Americas 40, and Oceania 14. https://simple.wikipedia.org/wiki/List_of_countries_by_continents

136 Gabriella Mulligan, *442 tech hubs active in Africa – GSMA* (Disrupt Africa, 22 Mar 2018) http://disrupt-africa.com/2018/03/442-tech-hubs-active-in-africa-gsma/

137 Extrapolated from: https://www.theatlas.com/charts/HyMDI_1ml and https://en.wikipedia.org/wiki/Lists_of_universities_and_colleges_by_country#Africa

138 For a list of cities over one million people, see: https://en.wikipedia.org/wiki/List_of_urban_agglomerations_in_Africa, for a list of cities larger than 300,000 people, see: United Nations World Urbanization Prospects 2014 by country: WUP2014-F12-Cities_Over_300K.xls

139 The following page lists over 100,000 NGOs in South Africa alone. See: Department of Social Development, South Africa, *2014/15 State of South African Registered Nonprofit Organizations* (Mar 2015) http://www.dsd.gov.za/npo/index2.php?option=com_docman&task=doc_view&gid=183&Itemid=116

140 To the degree that any of these crypto-tech and Blockchain solutions can deliver acceptable "censorship resistance" of security, either platform-level or the more preferred sovereign-level. "Atomic swaps" enable the use of different cryptocurrencies when two parties interact with each other, with the "exchange rate translation" being done automatically and directly on the level of the coin (or "atom"), not using an intermediary.

141 Over time, as Africa gets richer, increasing numbers of these customers will be African also.

142 https://www.fbi.gov/history/famous-cases/willie-sutton

143 Eric A. Posner, E. Glen Weyl, *Radical Markets: Uprooting Capitalism And Democracy For A Just Society* (Princeton, NJ: Princeton University Press, 2018)

144 This is a rough estimate. It includes Internet connectivity via

LEO micro satellites, connectivity, physical infrastructure including 5G and VR/XR, and the training of the people.

145 The stakeholders should include technical expertise in Blockchain (e.g. EOS/Block.one, Ethereum/EEF, Hyperledger), AI, genomics/data science, AR/XR, 5G, and IoT, and also cover contract negotiation with satellite and communication bandwidth providers.

146 Mutsa Chironga, Luis Cunha, Hilary De Grandis, Mayowa Kuyoro, "Roaring to life: Growth and innovation in African retail banking," p. 11 (McKinsey, Global Banking, February 2018) https://www.mckinsey.com/~/media/McKinsey/Industries/Financial%20Services/Our%20Insights/African%20retail%20bankings%20next%20growth%20frontier/Roaring-to-life-growth-and-innovation-in-African-retail-banking-web-final.ashx

147 Abdi Latif Dahir, *Mobile money is the key to growing Africa's banking sector* (Quartz Africa, 3 Apr 2018) https://qz.com/1243637/mobile-money-is-key-to-growing-banks-in-africa/ [8 Oct 2018]

148 Mutsa Chironga, Luis Cunha, Hilary De Grandis, Mayowa Kuyoro, "African retail banking's next growth frontier" (McKinsey Report, Feb 2018) https://www.mckinsey.com/industries/financial-services/our-insights/african-retail-bankings-next-growth-frontier

149 https://bitcoin.org/bitcoin.pdf

150 It would not even matter whether such a cryptocurrency would be totally new or pegged to a fiat currency as a so-called "stablecoin." A stablecoin pegged to the US Dollar would be an extension of the dollar, a stablecoin pegged to an African fiat currency would be rather unstable and weak.

151 Like the failed Petro of Venezuela. And who would want to be compared to that?

152 Apart from the "trillionaire" problem mentioned below, in the Epilogue of this book.

153 Maybe in form of a so-called "colored coin," which lets one add characteristics to specific coins that otherwise would be the same.

154 "Permissioned" means that coin governance is by invitation only,

while non-permissioned enables the public and therefore everybody to join in. Ripple and most Hyperledger implementations are examples of the first, Bitcoin and Ethereum are examples of the second.

155 A "genesis block" is the number of coins created at the moment when a new Blockchain platform is created. It typically is distributed to investors, individual founders, or whoever the founding community is.

156 DPoS works philosophically similar to an elected parliament to whose members citizens (in the Blockchain world: owners of coins) delegate the authority to make decisions on their behalf.

157 On a more basic level, IDG calls the platform used to enable such solutions in the first place a "Industry Cloud" or "Industry Collaborative Cloud." See: Eric Newmark, *Industry Cloud* (IDG, [no date]) https://www.idc.com/promo/thirdplatform/RESOURCES/ATTACHMENTS/Industry_Cloud_Q&A.pdf [22 Oct 2018]

158 See also: Erwin Rauch, Sven Seidenstricker, Patrick Dallasega, Robert Hämmerl, *Collaborative Cloud Manufacturing: Design of Business Model Innovations Enabled by Cyberphysical Systems in Distributed Manufacturing Systems* (Journal of Engineering, Volume 2016, Article ID 1308639) https://www.hindawi.com/journals/je/2016/1308639/ [22 Oct 2018]

159 Roslyn Layton, *5G Can Reinvent Ecommerce With VR and AR - If Policymakers Allow* (Forbes, 6 Jun 2018) https://www.forbes.com/sites/roslynlayton/2018/06/06/5g-can-reinvent-ecommerce-with-ar-and-vr-if-policymakers-allow/#7ffd22c63314 [8 Oct 2018]

160 Panos Constantinides, Ola Henfridsson, Geoffrey G. Parker, *Introduction—Platforms and Infrastructures in the Digital Age* (Information Systems Research 29(2):381-400, particularly 390ff) https://doi.org/10.1287/isre.2018.0794 [22 Oct 2018]

161 Research and Markets, *Future of African Backbone & Metro Fibre Networks, 2018* (22 Aug 2018) https://globenewswire.com/news-release/2018/08/22/1554923/0/en/Future-of-African-

Backbone-Metro-Fibre-Networks-2018-The-Need-to-Support-4G-and-FTTH-Network-Coverage-and-Prepare-for-5G-is-Transforming-Metro-Capacity-Requirements.html [18 Oct 2018]

162 The use of edge computing in combination with AI may in some areas eventually help alleviate the perception of this limitation somewhat, by performing calculations that enable a dramatic reduction of the transmitted data volume. Software may then trick the user into sensing an instantaneous connection that does in reality not exist.

163 Arbesman, Samuel, *The Half-Life of Facts – Why Everything We Know Has an Expiration Date* (New York: Penguin, 2013)

164 Ibid., page 32

165 https://www.technologyreview.com

166 Bitange Ndemo, Tim Weiss, A Call for Lifelong Learning Models in the Digital Age (Ottawa, Canada: IDRC – International Development Research Center, 2017? [no date]) http://fowigs.net/a-call-for-lifelong-learning-models-in-the-digital-age/ [22 Oct 2018]

167 K.N.C., G.S., P.K., *Daily Chart: Happy birthday world wide web – Society is adopting technology at an increasingly fast pace* (The Economist, 12 May 2014) https://www.economist.com/graphic-detail/2014/03/12/happy-birthday-world-wide-web?%3Ffsrc%3Dscn%2F=tw%2Fdc [8 Oct 2018]

168 Why a token? In theory, the token could be a currency, e.g. a dollar. But not everybody uses the same currency. Besides, the token is very specific. It is backed by marketable skills.

169 Erin Winick, *A new reports says AI will replace as many jobs as it kills – but take that with a huge grain of salt* (MIT Technology Review, 17 Jul 2018) https://www.technologyreview.com/the-download/611663/a-new-report-says-ai-will-replace-as-many-jobs-as-it-kills-take-that-with-a-huge/ [8 Oct 2018]

170 Lorraine Boissoneault, *Bismarck Tried to End Socialism's Grip – By Offering Government Healthcare* 14 Jul 2017) https://www.smithsonianmag.com/history/bismarck-tried-end-socialisms-grip-offering-government-healthcare-180964064/ [8 Oct 2018]

171 Yomi Kazeem, *Africans are living longer than at any point in the past 25 years* (Quartz Africa, 24 May 2016) https://qz.com/africa/690297/africans-are-living-longer-than-at-any-point-in-the-last-25-years/ [8 Oct 2018]

172 Fabian Urech, *Afrika lebt länger* (Neue Zürcher Zeitung, 4 Sep 2018) https://www.nzz.ch/international/afrika-lebt-laenger-ld.1417064 [18 Oct 2018]

173 World Health Organization, African Health Observatory, *Comprehensive Analytical Profile: WHO African Region (WHO)* http://aho.afro.who.int/profiles_information/index.php/AFRO:Index [8 Oct 2018]

174 Faras Ghani, *WHO: Africa's healthcare suffering from lack of funding* (Al Jazeera, 17 Dec 2017) https://www.aljazeera.com/news/2017/12/africa-healthcare-suffering-lack-funding-171217122635626.html [8 Oct 2018]

175 Petrie Jansen van Vuuren, *FACTSHEET: Africa's leading causes of death* (AfricaCheck, 14 Aug 2017, based on data from the World Health Organization), https://africacheck.org/factsheets/factsheet-africas-leading-causes-death/ [18 Oct 2018]

176 Daniel Armah-Attoh, Edem E. Selormey, Richard Houessou, *More Africans are accessing health care – but states still have work to do* (The Conversation, 25 May 2016) http://theconversation.com/more-africans-are-accessing-health-care-but-states-still-have-work-to-do-59767 [8 Oct 2018]

177 Laurant Musango, Riku Elovainio, Juliet Nabyonga, Bokar Toure (WHO), *The state of health financing in the African Region* (WHO, African Health Observatory, Issue $16) http://www.aho.afro.who.int/en/ahm/issue/16/reports/state-health-financing-african-region [8 Oct 2018]

178 The World Bank, *GDP per capita (current US$)* https://data.worldbank.org/indicator/NY.GDP.PCAP.CD?locations=ZG [8 Oct 2018]

179 I got a deeper appreciation of this general management concept when reading *Thinking in Bets* by Annie Duke (New York: Penguin Random House, 2018), pp. 219-221.

180 For 2016, the total federal tax base was $3.3 trillion dollars,

which is about the same number as the total "national health expenditure." https://www.cbo.gov/publication/52408 and https://www.cms.gov/research-statistics-data-and-systems/statistics-trends-and-reports/nationalhealthexpenddata/nhe-fact-sheet.html [13 Oct 2018]

181 Storage of a nano-chip right below people's skin has been successfully tested before, as has the storing of data inside DNA. See: Yael Grauer, *A practical guide to microchip implants* (Ars Technica, 3 Jan 2018) https://arstechnica.com/features/2018/01/a-practical-guide-to-microchip-implants/ [18 Oct 2018]

182 David Constantine, *Sia Vs. Storj Vs. Amazon – Price Compare* (Steemit, January 2018) https://steemit.com/sia/@davidconstantine/sia-vs-storj-vs-amazon-price-compare [13 Oct 2018]

183 As of July 2017, the cost was estimated to be $1,121. Wetterstrand KA, *DNA Sequencing Costs: Data from the NHGRI Genome Sequencing Program (GSP)* Available at: www.genome.gov/sequencingcostsdata [13 Oct 2018]

184 As of mid-2018, many so-called "high-throughput" gene-sequencing techniques also used chemical "reagents," constituting another significant cost component.

185 This assumes an average cost of about $800 to $1000 for whole genome sequencing per person.

186 Based on rough calculations adding up the required capabilities for the currently most ambitious projects. See: Labiotech.eu, *The Past, Present and Future of Genome Sequencing* (Labiotech.eu, 4 Sep 2018) https://labiotech.eu/features/genome-sequencing-review-projects/

187 This requires approaches for privacy-preserving data analytics, including but not limited to so-called homomorphic encryption, which enables the analysis of encrypted data.

188 Vitalik Buterin, *The Meaning of Decentralization* (Medium: 6 Feb 2017) https://medium.com/@VitalikButerin/the-meaning-of-decentralization-a0c92b76a274 [8 Oct 2018]

189 International Finance Corporation, *IFC In The Power Sector*

(World Bank Group) https://www.ifc.org/wps/wcm/connect/b82f23fc-6285-43ef-9cd5-938f725fe604/PowerInfographic.pdf?MOD=AJPERES [8 Oct 2018]

190 Irene Sun, *How can Blockchain Change the Energy Industry?* (Medium, 25 Mar 2018) https://medium.com/kiwi-new-energy/how-can-blockchain-change-the-energy-industry-84e8f1db4c90 [8 Oct 2018]

191 Ben Schiller, *This Blockchain-Based Energy Platform Is Building A Peer-to-Peer Grid* (Fast Company, 16 Oct 2017) https://www.fastcompany.com/40479952/this-blockchain-based-energy-platform-is-building-a-peer-to-peer-grid [8 Oct 2018]

192 Nikolaj Martyniuk, *Blockchain Establishes Decentralized Energy Market* (Medium, WePower, 11 Mar 2018) https://medium.com/wepower/blockchain-establishes-decentralized-energy-market-ae413952f6da [8 Oct 2018]

193 RFID stands for "radio-frequency identification." NFC is the acronym for "near-field communication."

194 Forbes 2018 ranking shows Donald Trump as number 766 with $3.1 billion, compared to Jeff Bezos' $112 billion. At the time of this writing the numbers were rank 723 and $3.1 billion for Donald Trump, and $142.3 billion for Jeff Bezos. See: *Forbes, The World's Billionaires, 2018 Ranking* (Forbes, 23 Oct 2018) https://www.forbes.com/billionaires/list/18/#version:static [23 Oct 2018]

195 Cherry Reynard, *Who are the richest cryptocurrency investors?* (The Telegraph, 25 May 2018) https://www.telegraph.co.uk/technology/digital-money/richest-crypto-investors/ [11 Oct 2018]

196 Pamela Ambler, Angel Au-Yeung, Grace Chung, Jeff Kauflin, Alex Konrad, Laura Shin, Nathan Vardi, *The Richest People in Cryptocurrency* (Forbes, 6 Feb 2018) https://www.forbes.com/richest-in-cryptocurrency/#ce77cc41d496 [11 Oct 2018]

197 Altcoins in this context is the term describing all Blockchain or cryptocurrencies that are not Bitcoin. As of September 2018

largest ones were Ethereum, Ripple, EOS, and Stellar.

198 The BTC camp often describes its main value as a "store of value," while the BCH camp states that they want to be electronic cash for the world.

199 Sharding is a technology approach to Ethereum with the goal to increase performance.

200 Yuval Noah Harari, *Are we about to witness the most unequal societies in history?* (The Guardian, 24 May 2018) https://www.theguardian.com/inequality/2017/may/24/are-we-about-to-witness-the-most-unequal-societies-in-history-yuval-noah-harari [8 Oct 2017]

201 Ibid.

Made in United States
North Haven, CT
10 November 2026

22081996R00152